Methods in Molecular Biology

Series Editor
John M. Walker
School of Life and Medical Sciences
University of Hertfordshire
Hatfield, Hertfordshire, AL10 9AB, UK

For further volumes:
http://www.springer.com/series/7651

Superantigens

Methods and Protocols

Edited by

Amanda J. Brosnahan

Science Department, Concordia University - Saint Paul, St. Paul, MN, USA

Humana Press

Editor
Amanda J. Brosnahan
Science Department
Concordia University–Saint Paul
St. Paul, MN, USA

ISSN 1064-3745 ISSN 1940-6029 (electronic)
Methods in Molecular Biology
ISBN 978-1-4939-8026-0 ISBN 978-1-4939-3344-0 (eBook)
DOI 10.1007/978-1-4939-3344-0

Springer New York Heidelberg Dordrecht London

Softcover re-print of the Hardcover 1st edition 2016

Printed on acid-free paper

Humana Press is a brand of Springer
Springer Science+Business Media LLC New York is part of Springer Science+Business Media (www.springer.com)

Preface

The bacterial exotoxins made by *Staphylococcus aureus* and *Streptococcus pyogenes* were officially termed superantigens in 1989 by Phillipa Marrack and John Kappler due to their unique ability to stimulate large numbers of T cells compared to "normal" antigens. Since then, the field of superantigen research has dramatically increased, as is evident by the over 5000 publications referencing the term "superantigen" in the PubMed.gov database. These publications have examined the structure and function of superantigens, interactions of superantigens with other host cells, and the role of superantigens in diseases as varied as menstrual toxic shock syndrome to necrotizing pneumonia to atopic dermatitis. The purpose of *Superantigens* is to combine protocols that encompass the true variety of investigations done on superantigens in the fields of microbiology, immunology, molecular biology, biochemistry, and cellular biology, with a strong focus on disease models utilized to determine the role of superantigens in human disease.

St. Paul, Minnesota, USA *Amanda J. Brosnahan*

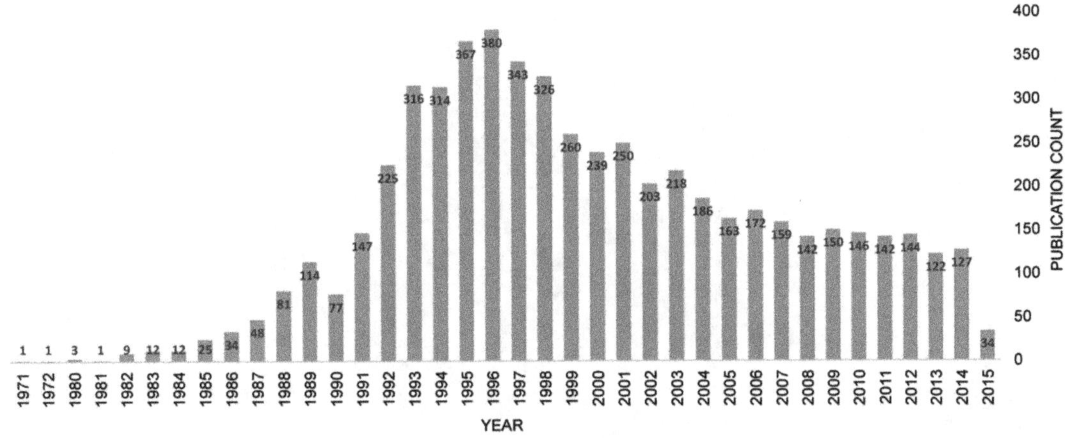

Number of publications by year when searching for "superantigen" in the PubMed database at: PubMed [Internet]. Bethesda (MD): National Library of Medicine (US). [1946] to [2015 June 11]. Available from: http://pubmed.gov/

Contents

Contributors

RANU AGRAWAL • *Biotechnology Division, Defense R&D Establishment, Gwalior, India; Department of Chemistry, Chaudhary Charan Singh University, Meerut, India*

KENNETH L. BOST • *Department of Biological Sciences, University of North Carolina at Charlotte, Charlotte, NC, USA*

LAURA M. BRESHEARS • *Department of Experimental and Clinical Pharmacology, College of Pharmacy, University of Minnesota, Minneapolis, MN, USA*

AMANDA J. BROSNAHAN • *Department of Science, Concordia University–Saint Paul, St. Paul, MN, USA*

STEFANO CASERTA • *Division of Clinical Medicine, BSMS—Brighton and Sussex Medical School, University of Sussex, Falmer, UK*

YU-CHENG CHIANG • *Department of Food Science and Technology, Hung Kuang University, Taichung, Taiwan*

VAIDEHI CHOWDHARY • *Division of Rheumatology, Department of Medicine, Mayo Clinic, Rochester, MN, USA*

E. MICHAEL DANIELSEN • *Faculty of Health Sciences, Department of Cellular and Molecular Medicine, The Panum Institute, University of Copenhagen, Copenhagen N, Denmark*

S.M. MANSOUR HAERYFAR • *Department of Microbiology and Immunology, Schulich School of Medicine and Dentistry, Western University, London, ON, Canada; Lawson Health Research Institute, London, ON, Canada*

GERT H. HANSEN • *Faculty of Health Sciences, Department of Cellular and Molecular Medicine, The Panum Institute, University of Copenhagen, Copenhagen N, Denmark*

D.V. KAMBOJ • *Biotechnology Division, Defense R&D Establishment, Gwalior, India*

KATHERINE J. KASPER • *Department of Microbiology and Immunology, Schulich School of Medicine and Dentistry, Western University, London, ON, Canada*

ASHTON L. KROGMAN • *Department of Immunology, Mayo Clinic, Rochester, MN, USA*

SHENG-CHIH LI • *Department of Food Science and Biotechnology, National Chung-Hsing University, Taichung, Taiwan*

KARIN LINDKVIST-PETERSSON • *Department of Experimental Medical Science, BMC C 13, Lund University, Lund, Sweden*

MARTIN J. LLEWELYN • *Division of Clinical Medicine, BSMS—Brighton and Sussex Medical School, University of Sussex, Falmer, UK; Brighton and Sussex University Hospitals NHS Trust, Brighton, UK*

ULVER SPANGSBERG LORENZEN • *Faculty of Health Sciences, Department of Cellular and Molecular Medicine, The Panum Institute, University of Copenhagen, Copenhagen N, Denmark*

MARY J. MANTZ • *Dows Institute for Dental Research, College of Dentistry, University of Iowa, Iowa City, IA, USA*

JOHN K. MCCORMICK • *Department of Microbiology and Immunology, Schulich School of Medicine and Dentistry, Western University, London, ON, Canada; Lawson Health Research Institute, London, ON, Canada*

JOSEPH A. MERRIMAN • *Department of Microbiology, University of Iowa, Iowa City, IA, USA*
JACK ODLE • *Department of Animal Science, North Carolina State University, Raleigh, NC, USA*
JOO YOUN PARK • *Department of Basic Sciences, College of Veterinary Medicine, Mississippi State University, Mississippi State, MS, USA*
MARNIE L. PETERSON • *Department of Experimental and Clinical Pharmacology, College of Pharmacy, University of Minnesota, Minneapolis, MN, USA*
KENNETH J. PILLER • *Department of Biological Sciences, University of North Carolina at Charlotte, Charlotte, NC, USA*
GOVINDARAJAN RAJAGOPALAN • *Department of Immunology, Mayo Clinic, Rochester, MN, USA; Division of Infectious Diseases, Department of Medicine, Mayo Clinic, Rochester, MN, USA*
KARIN E.J. RÖDSTRÖM • *Department of Experimental Medical Science, BMC C13, Lund University, Lund, Sweden*
JESSICA H. ROTSCHAFER • *Department of Microbiology, University of Minnesota, Medical School, Minneapolis, MN, USA*
WILMARA SALGADO-PABÓN • *Department of Microbiology, University of Iowa, Iowa City, IA, USA*
PATRICK M. SCHLIEVERT • *Department of Microbiology, University of Iowa, Iowa City, IA, USA*
KEUN SEOK SEO • *Department of Basic Sciences, College of Veterinary Medicine, Mississippi State University, Mississippi State, MS, USA*
LOKENDRA SINGH • *Biotechnology Division, Defense R&D Establishment, Gwalior, India*
PAWAN KUMAR SINGH • *Department of Ophthalmology, Anatomy and Cell Biology, Wayne State University School of Medicine, Detroit, MI, USA*
CHRISTOPHER A. SQUIER • *Dows Institute for Dental Research, College of Dentistry, University of Iowa, Iowa City, IA, USA*
CHRISTOPHER S. STACH • *Department of Microbiology, University of Iowa, Iowa City, IA, USA*
CHAD H. STAHL • *Department of Animal Science, North Carolina State University, Raleigh, NC, USA*
KRISTI L. STRANDBERG • *Department of Microbiology, University of Minnesota Medical School, Minneapolis, MN, USA*
AMANDA L. TAYLOR • *Division of Clinical Medicine, BSMS—Brighton and Sussex Medical School, University of Sussex, Falmer, UK*
NADIA TERRAZZINI • *Pharmacy and Biomolecular Sciences, University of Brighton Moulsecoomb, Brighton, UK*
SHUO-WEN TSAI • *Department of Food Science and Biotechnology, National Chung-Hsing University, Taichung, Taiwan*
HAU-YANG TSEN • *Department of Food Science and Technology, Hung Kuang University, Taichung, Taiwan*
ADRIENNE T. WAKABAYASHI • *Department of Microbiology and Immunology, Schulich School of Medicine and Dentistry, Western University, London, ON, Canada*
STACEY X. XU • *Department of Microbiology and Immunology, Schulich School of Medicine and Dentistry, Western University, London, ON, Canada*
JOSEPH J. ZEPPA • *Department of Microbiology and Immunology, Schulich School of Medicine and Dentistry, Western University, London, ON, Canada*

Chapter 1

Animal Models Used to Study Superantigen-Mediated Diseases

Amanda J. Brosnahan

Abstract

Superantigens secreted by *Staphylococcus aureus* and *Streptococcus pyogenes* interact with the T-cell receptor and major histocompatibility class II molecules on antigen-presenting cells to elicit a massive cytokine release and activation of T cells in higher numbers than that seen with ordinary antigens. Because of this unique ability, superantigens have been implicated as etiological agents for many different types of diseases, including toxic shock syndrome, infective endocarditis, pneumonia, and inflammatory skin diseases. This review covers the main animal models that have been developed in order to identify the roles of superantigens in human disease.

Key words Superantigen, *Staphylococcus aureus*, *Streptococcus pyogenes*, Toxic shock syndrome, Animal model, Pneumonia, Endocarditis, Atopic dermatitis, Emesis

1 Discovery of Superantigens

Although a variety of exotoxins produced by the gram-positive bacterial pathogens *Staphylococcus aureus* and *Streptococcus pyogenes* had been discovered earlier, it was not until 1989 that a group of these toxins would be named superantigens by Marrack and Kappler due to their unique ability to stimulate larger numbers of T cells compared to a "typical" antigen [1]. The superantigens bind to the variable region of the β chain of the T-cell receptor (Vβ-TCR) on T cells and major histocompatibility complex II (MHC II) on antigen-presenting cells, such as macrophages [2–6]. This interaction often leads to activation of approximately 20 % of all T cells and release of cytokines and chemokines from both cell types. The hallmark identification of potential superantigens involves a lymphocyte mitogenicity assay in which increasing concentrations of the potential superantigen are incubated with peripheral blood mononuclear cells (PBMCs). Tritiated thymidine

Amanda J. Brosnahan (ed.), *Superantigens: Methods and Protocols*, Methods in Molecular Biology, vol. 1396, DOI 10.1007/978-1-4939-3344-0_1, © Springer Science+Business Media New York 2016

is added to the cultures after a few days to see if activated and dividing cells are present as dividing cells will incorporate the radiolabel into their DNA [7]. This technique is described in Chapter 2 of this volume.

The staphylococcal superantigens include the staphylococcal enterotoxins (SE) A, B, C, D, E and I, and SE-like (SEl) superantigens, including SEl G, H, J–X [8–10]. The enterotoxins are known to cause emesis, while the SE-like toxins have either not been tested in a primate model of emesis or have been found to be non-emetic. There is no SE or SEl F because this toxin was renamed toxic shock syndrome toxin-1 (TSST-1), the final superantigen (to date) in the staphylococcal group. The streptococcal superantigens made by *S. pyogenes* (also known as group A *Streptococcus*—GAS) include the streptococcal pyrogenic exotoxins (SPE) A, C, G, H, I, J, K, L, M, streptococcal superantigen (SSA), and streptococcal mitogenic exotoxin Z (SMEZ) [11]. Recently, three more superantigen genes have been identified in group C and group G *Streptococcus*: originally designated *szeN*, *szeP*, and *szeF*, but proposed by Commons et al. to be renamed SPE N, SPE P, and SPE F due to their high similarity to the GAS superantigens [11, 12]. This review will focus on superantigens made by *S. aureus* and *S. pyogenes*, however other superantigen proteins have been discovered in *Mycoplasma arthritidis* [13] and *Yersinia* species [14–16].

2 Structure of Superantigens

Superantigens are 19–30 kDa weight proteins that consist of two distinct globular domains. The N-terminus contains an oligosaccharide/oligonucleotide (OB) fold, while the C-terminus contains a β-grasp domain. Superantigens can bind MHC II through a low-affinity site in the OB fold and/or a high-affinity site in the β-grasp domain and this interaction may be antigen peptide-dependent or independent, depending on the superantigen [17]. Slightly different structural groupings based on MHC II binding and amino acid sequences have been proposed by Fraser and Proft [17] and Schlievert and Bohach [18]. For example, TSST-1 binds to the alpha chain of MHC II in a peptide-dependent manner and it has a unique amino acid sequence compared to the rest of the superantigens, so it tends to be grouped on its own [19]. The region for TCR binding resides in a groove between both major folds. Figure 1 shows two examples of how superantigens are thought to interact with MHC II and TCR receptors: SPE A binds the two receptors toward the side of the complex, as a wedge, whereas SPE C binds between the two receptors, interacting with the high-affinity beta chain of MHC II, to appear as three beads on a string.

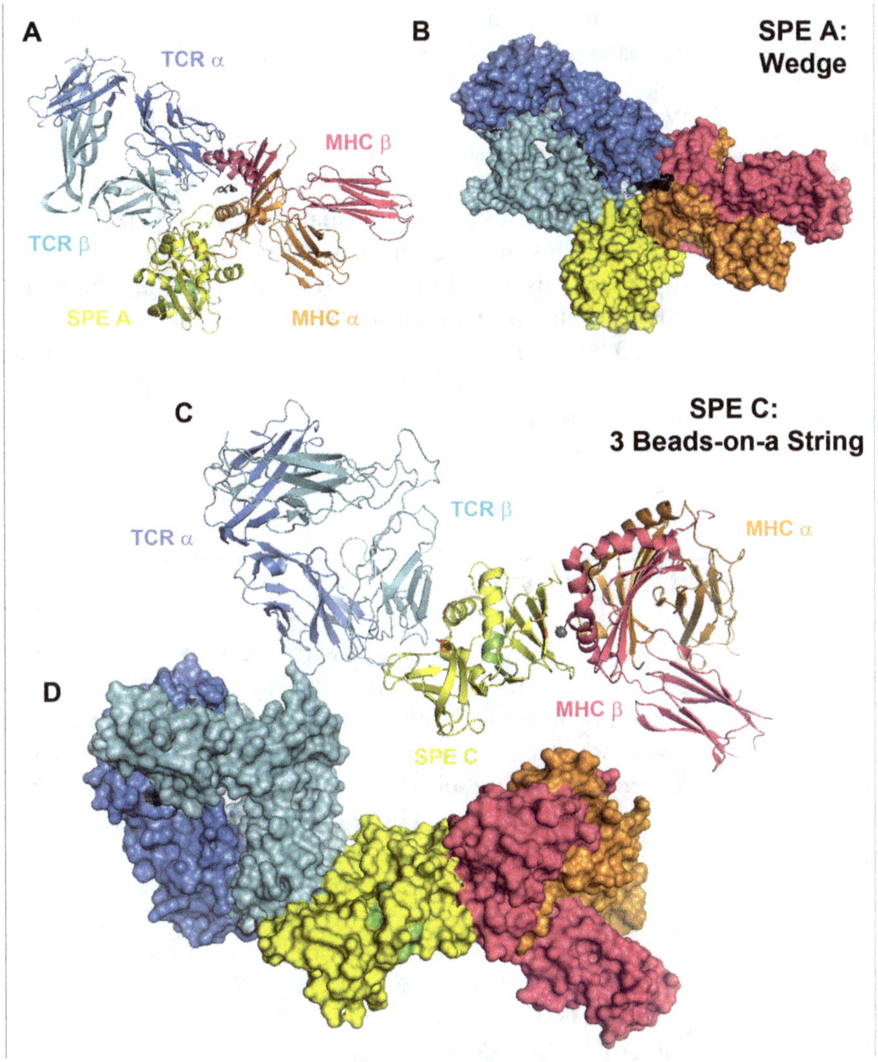

Fig. 1 Superantigen interactions with TCR and MHC II. Ternary structure models of SPE A and SPE C interacting with TCR and MHC II molecules. (**a**) *Cartoon* and (**b**) surface views of SPE A ternary complex, based on the following binary complexes: SPE A/TCR-β (PDB 1L0Y), TCR-αβ (PDB 1J8H) and SEB /MHC II (PDB 1SEB). (**c**) *Cartoon* and (**d**) surface views of SPE C ternary complex, based on the following binary complexes: SPE C/TCR-β (PDB 1KTK), TCR-αβ (PDB 1J8H) and SPE C-MHC II (PDB 1HQR). Two main models of superantigen interactions are shown: the wedge (SPE A) and three beads on a string (SPE C). Superantigens are shown in *yellow*, with residues critical for MHC II or TCR binding shown in *red* and the epithelial binding region highlighted in *green*. The epithelial binding site remains distinct from the MHC II and TCR binding sites and is not thought to be involved in this interaction. TCR α-chains are shown in *dark blue*, TCR β-chains are shown in *light blue*, MHC II α-chains are shown in *orange*, and MHC II β-chains are shown in *pink*. A zinc ion is shown as a *green ball* in the SPE C ternary complex, while peptides in the MHC II groove are shown in *black* in all complexes. Images generated using PYMOL (DeLano Scientific LLC, South San Francisco, CA, USA). Reprinted with permission from FEBS Journal © 2011 [22]

Distinct from the MHC II and TCR binding sites on superantigens is a dodecapeptide (12-amino acid) region that has been implicated in interactions with epithelial cells (*see* Fig. 1) [20–22]. Work done by Shupp et al. originally implicated this region as required for transcytosis across intestinal epithelial cells [20], but this region has also been shown to induce proinflammatory responses from human vaginal epithelial cells [22, 23]. More recently, this interaction has been attributed to the ability of superantigens to bind CD40 on human vaginal epithelial cells through the dodecapeptide region. It was also postulated that CD40 binding on B cells in vivo may contribute to adjuvant effects that can be seen with certain superantigen mutants [24]. The same dodecapeptide region has been implicated in binding of superantigens to the costimulatory molecule on T cells, CD28, as part of the MHC II–TCR interaction [25, 26].

3 Diseases Associated with Superantigens

The main disease associated with the staphylococcal and streptococcal superantigens is toxic shock syndrome (TSS). TSS is characterized by fever, rash, hypotension, and involvement of three organ systems, as well as peeling of the skin upon recovery. TSS caused by *S. aureus* can be either menstrual-associated (mTSS) or non-menstrual-associated. Those cases of mTSS usually occur within 2 days of menstruation and are often associated with tampon use [27, 28]. The association with tampons is thought to be due to the upregulation of superantigen production by *S. aureus* in the presence of oxygen, which is introduced into the normally anaerobic environment by tampons [29–31]. TSST-1 is the major superantigen associated with all cases of mTSS, dating back to the original epidemic of mTSS in the early 1980s [32, 33]. Non-menstrual TSS is primarily caused by TSST-1 (50 %) and SEB and SEC (50 % together), which are produced in higher concentrations than other superantigens (up to 10^6-fold more) [34]. Non-menstrual TSS can occur in many different forms, including the more severe presentations of purpura fulminans and necrotizing fasciitis [35–37], and can be subsequent to other infections, such as pneumonia. The overall incidence of both mTSS and TSS is 1–3 per 100,000 population with a case:fatality rate of 5–10 % [22].

Although the main symptoms of streptococcal TSS (STSS) are very similar to staphylococcal TSS, there are some key differences. STSS is not menstrual-associated and it usually occurs in the context of bacteremia, unlike staphylococcal TSS, in which *S. aureus* remains localized on a mucosal surface. STSS is also often associated with necrotizing fasciitis or myositis, whereas those presentations are less common in staphylococcal TSS cases [22]. The major toxin initially associated with STSS was SPE A [38], which has historically been implicated in scarlet fever, another disease caused by GAS. In fact, some more severe forms of scarlet fever may actually have been STSS. Other streptococcal superantigens such as SMEZ have more recently been implicated in severe streptococcal disease, including STSS (reviewed in ref. 11).

Other diseases have also been associated with superantigens. The staphylococcal enterotoxins are implicated in staphylococcal food poisoning [39] and pseudomembranous enterocolitis [40]. Multiple staphylococcal superantigens have been implicated in atopic dermatitis [41], whereas both streptococcal and staphylococcal superantigens have been associated with psoriasis [42–44]. More obscure syndromes, such as staphylococcal extreme pyrexia [45] and pediatric autoimmune neuropsychiatric disorders associated with streptococcal infections (PANDAS) [46] have also been linked to superantigens. Most recently, a link between type II diabetes and staphylococcal superantigens has been proposed [47–49].

4 Animal Models to Study Superantigen-Mediated Diseases

The remainder of this review will focus on the main animal models that have been used to study how superantigens contribute to various diseases. All animal models discussed are listed in Table 1.

4.1 Toxic Shock Syndrome Models

Since superantigens are predominantly associated with TSS, most animal models are centered on mimicking the symptoms of this syndrome. Two key review articles were published in the late 1980s to assess the main animal models available at that time—one by de Azavedo and another by Quimby and Nguyen [50, 51]. Two animal models were identified at that point: the rabbit and the baboon. The baboon model was put forward as the better model by Quimby and Nguyen, due to work done in their lab that indicated it provided a response that most closely mimicked the symptoms seen in human cases of TSS. Intravenous injection of purified TSST-1, albeit in higher doses than that required for rabbits, can lead to symptoms of TSS in baboons and monkeys [51, 52], however intravaginal administration of TSST-1-producing bacterial strains in baboons and monkeys did not lead to TSS [50, 51, 53]. Very little work has been done using baboons or monkeys to mimic TSS since these early reports, presumably because models that were cheaper and easier to work with became readily available.

The rabbit model has been used extensively since it was first put forward in the 1980s. Rabbits are generally easy to handle, are non-aggressive, and are readily available and economical compared to larger research animals [54]. There are a number of variations on rabbit models of TSS, including ones that require lipopolysaccharide (LPS) and ones that do not. Additionally, there are multiple

Table 1
Summary of animal models used to study superantigen-mediated diseases

Animal	Disease	Model types	Reference(s)
Rabbit	TSS	LPS enhancement	[10, 23, 55–57]
		Small-dose diffusion	[58–63]
	Infective endocarditis	IV after mechanical heart valve damage	[85]
	Pneumonia	Intrabronchial inoculation	[86, 88]
Mouse	TSS	D-galactosamine	[68–70]
		Actinomycin D	[67]
		Superantigen-only (intransal and intraperitoneal)	[74, 75]
		MHC class II transgenic	[73, 76, 77, 79, 80]
	Pneumonia	Intratracheal inoculation of MHC class II transgenic	[87]
	Atopic dermatitis	BALB/c—intradermal and epicutaneous	[99–101]
		SCID mice reconstituted with human PBMCs	[102]
	Psoriasis	SCID mice transplanted with skin from humans	[43, 44, 104, 105]
Non-human primate	TSS	IV; baboons and monkeys	[51, 52]
		Intravaginal; baboons and monkeys	[50, 51, 53]
	Emesis	Oral	[90, 91]
		IV	[51]
Pig	TSS	Subcutaneous	[82]
		IV	[83, 84]
	Emesis	Oral	[92]
		IV	[83, 93]
Ferret	Emesis	Oral	[94]
		IV	[95]
House musk shrew	Emesis	Oral	[96, 97]
		Intraperitoneal	[98]

TSS toxic shock syndrome, *IV* intravenous, *MHC* major histocompatibility complex

methods of superantigen exposure including intravenous exposures, vaginal exposures that simulate menstrual TSS, and various small-dose diffusion models. As outlined by de Azavedo in her 1989 review, rabbits display all of the hallmark features of TSS, except for the rash that is followed by desquamation (some rabbits do develop a reddening of the skin) and vomiting (which rabbits are incapable of doing) [50].

One of the most commonly used rabbit models of TSS is one involving LPS enhancement, in which small doses of superantigen are administered prior to administration of a sublethal dose of LPS. This model was originally described by Kim and Watson [55] for streptococcal superantigens and then by Schlievert [56] for staphylococcal superantigens. The model tests the ability of a superantigen to enhance the effects of LPS, usually leading to signs of TSS and eventual death. The timing and doses depend on the superantigen being tested, but typically a superantigen is given 3–4 h prior to a low dose of LPS (1–10 μg/kg; LD_{50} for LPS alone is 500 μg/kg) and rabbits are monitored for up to 48 h for signs of TSS or death. This allows researchers to administer a smaller dose of superantigen to the rabbits as well, since the LD_{50} of some superantigens can be as high as 3500 μg/kg [55]. Conversely, administering the LPS first or simultaneously with the superantigen does not lead to TSS [55, 56]. This model was originally used with intravenous administration of superantigen, but has been adapted to be used with vaginal [10, 23, 57] and oral [57] administration of superantigen as well. Details of the LPS-enhancement model can be seen in Chapter 5 of this volume.

Rabbit models that do not require the use of LPS include those that administer superantigen or superantigen-producing bacterial strains to the animals through the use of a diffusion chamber, such as a perforated golf (whiffle) ball implanted into the side of the rabbit. The polyethylene whiffle balls are subcutaneously placed via surgery and rabbits are allowed to heal for a period of time prior to administration of the superantigen or bacterial challenge [58–60]. Alternatively, mini-osmotic pumps that are designed to exchange fluid can be preloaded with superantigen and inserted into small incisions on the rabbit's side [61–63]. The slow, continuous infusion of superantigen appears to mimic the symptoms of TSS in the absence of exogenous LPS, albeit it takes longer for this development to occur than the LPS-enhancement model described above and it may require a higher overall dose of superantigen.

Mice and other rodents appear to be resistant to the effects of superantigens [50, 51], and we now know that this is due to the slight difference in HLA genotypes between susceptible humans and non-susceptible rodents. When mice models were "humanized" with human-specific HLA genotypes, the transgenic mice were found to be much more susceptible to superantigen-mediated TSS [64–66]. Nonetheless, other murine models of superantigen-mediated TSS do exist, some of which require another sensitizing agent such as actinomycin D or D-galactosamine. The actinomycin D model relies on ability of actinomycin D to enhance the effects of tumor necrosis factor (TNF), which is a key player in the development of TSS [67]. Non-humanized BALB/cJ mice, normally resistant to SEB up to 500 μg per mouse, became sensitized when pre-treated with actinomycin D, although higher doses of actinomycin D (20 μg/mouse) and SEB (50–100 μg/mouse) were required to elicit lethality in this model [67]. Likewise, galactosamine murine models rely on damage done to hepatocytes such that the mice become increasingly sensitive to LPS [68, 69]. Treating C57BL/6 mice with 300 mg/kg galactosamine allowed for 100 % lethality to just 0.01 μg/mouse LPS, whereas in the absence of galactosamine, the LD_{50} of LPS is over 100 μg [68]. Meithke et al. went on to show that SEB (20 μg/mouse) could show the same enhancement after pre-treatment of BALB/c, C3H, and C57BL/6 mice with D-galactosamine (20 mg/mouse), in the absence of exogenous LPS [70]. The hepatic injury induced by D-galactosamine is extremely toxic such that it has been used to study viral hepatitis [71, 72], causing some researchers to question how accurate the murine model is in the context of human TSS [73].

Other murine models do exist that appear to cause a superantigen-mediated toxicity in the absence of potentiating agents. For example, there is a model that utilizes a dual dose of superantigen to elicit lethality in non-humanized mice. This superantigen-only model utilizes 5 μg of SEB administered intranasally followed by 2 μg of SEB given intraperitoneally 2 h later to C3H/HeJ mice [74, 75]. This model caused 100 % lethality by 120 h in 70 mice tested. These mice also experienced a temperature drop compared to controls, which the researchers concluded indicated that the mice were undergoing shock comparable to that seen with other murine models [74]. Monocyte chemoattractant protein (MCP)-1 was found in the lungs of these mice and it is postulated that the high levels of MCP-1 induced in this model may recruit enough macrophages to the lungs to properly initiate toxemia.

Mice transgenic for human HLA genotypes have more recently been used to further elucidate the role of superantigens in TSS, as well as other forms of severe staphylococcal and streptococcal disease. The HLA class II genes in humans are HLA-DR, -DQ, and -DP, and they are extremely polymorphic [76]. In fact, certain class II HLA haplotypes (DRB1*15/DQB1*06) have been associated with increased protection from developing severe streptococcal diseases, based on the lowered cytokine response of those haplotypes to superantigens [77, 78]. The use of human class II transgenic mice has continued to expand, allowing researchers to examine the role of superantigens in other areas of staphylococcal and streptococcal disease. For example, *S. pyogenes* can initiate an infection in the nasopharynx of class II transgenic mice due to the ability of its superantigens to interact with the human MHC II in the transgenic mouse compared to a non-transgenic mouse [79]. On the other end of things, vaginally administered SEB was shown to induce systemic inflammation in class II transgenic mice, indicating a role for SEB in menstrual TSS [80]. One limitation of the class II transgenic (humanized) mouse model is that due to the extreme polymorphism of the HLA genotype, it is impossible to have a mouse strain (or strains) that adequately covers every potential human haplotype, thus each of these studies only provide information on a specific haplotype [81]. Detailed explanations of the humanized mouse model for superantigen-mediated diseases are given in Chapters 8 and 9 of this volume.

Pigs have also been used to characterize the response to superantigens if injected subcutaneously or intravenously. Minnesota miniature breed suckling piglets (14 days old) and adult pigs developed some signs of TSS (increased temperature and organ dysfunction) when given 100 μg/kg TSST-1 subcutaneously, whereas germ-free piglets did not [82]. None of the pigs died in this study, but it was only carried out for 24 h so it is possible that lethality may have occurred upon later time points. Fourteen-day-old Yorkshire piglets were used to show that intravenous administration of SEB at a lethal dose (150 l g/kg) could elicit symptoms of TSS [83]. Increased body temperatures, diarrhea, anorexia, and shock developed prior to death in a biphasic response that is similar to that seen in humans. The main difference that the researchers noted is that the majority of the edema seen was in the abdomen, as opposed to the pulmonary edema that is seen in humans and primates, but other shock-like diseases in pigs also lead to abdominal edema, indicating that it may be a species difference [83].

A sublethal superantigen challenge model has similarly been developed in pigs [84]. Piglets infected intravenously with just 25 μg/kg of SEB led to incapacitation, with symptoms of lethargy, emesis and/or diarrhea, and prostration. These authors argued that animal models that require higher toxin doses and lead to lethality do not adequately allow for assessment of bioterrorism events in which incapacitation, and not lethality, is the endpoint. Additionally, cytokine responses can be analyzed in this model more readily than in rabbits as more research tools have been developed for pigs than for rabbits at this point in time and the authors argue that the cytokine response in pigs may be more physiologically relevant than that seen in a genetically altered mouse. A more detailed description of this sublethal porcine model is given in Chapter 10 of this volume.

4.2 Cardiopulmonary Disease Models

The role of superantigens in severe staphylococcal pneumonia and infective endocarditis has been studied using rabbit and HLA class II transgenic mouse models [85–87]. Intrabronchial inoculation of superantigen-producing strains of *S. aureus* or purified superantigens into rabbits was used to show that superantigens are at least partially responsible for lethal pulmonary infections [86]. Immunization against superantigens in this model has prevented death upon challenge with *S. aureus* or superantigens [86, 88]. Specific details about the pulmonary rabbit model can be found in Chapter 7 of this volume. HLA class II transgenic mice undergoing intratracheal inoculation with *S. aureus* strains that produced SEB were more likely to die than those inoculated with an isogenic SEB(-) strain, thus providing another basic model for examining the role of superantigens in pulmonary infections [87].

S. aureus is the leading cause of infective endocarditis, and this disease manifestation can be seen as a complication of bacterial sepsis [89]. Rabbits have been used to demonstrate the role of superantigens in the development of infective endocarditis [85]. Using a hard plastic catheter, researchers are able to sufficiently damage the heart valve, at which point the catheter can be removed and the rabbit can be sutured up. Rabbits are then intravenously injected with *S. aureus* strains that do or do not produce superantigens. At the conclusion of the experiment, hearts can be examined for vegetations of bacteria at the site of valve injury. Using this model, Salgado-Pabón et al. have shown that a community-associated, methicillin-resistant strain (MW2) that produces SEC caused lethality in 80 % of the infected rabbits, whereas an isogenic SEC(-) strain did not [85]. A more detailed explanation of this model can be found in Chapter 6 of this volume.

4.3 Emesis Models

Before they were characterized as superantigens, the SEs were studied for their role in staphylococcal food poisoning. Emesis has predominantly been studied in non-human primate models, however a few studies examining emesis in house musk shrews, ferrets, and pigs have also been conducted. The gold standard monkey feeding assay for staphylococcal enterotoxins was published by Bergdoll in 1988, although experiments with staphylococcal strains associated with food poisoning had been conducted as early as the 1930s [90]. In the rhesus monkey feeding assay, which was developed based on earlier studies [91], approximately 50 ml of staphylococcal culture fluid with presumed enterotoxins (or purified enterotoxin) are administered intragastrically by a stomach tube and animals are monitored for vomiting for 5 h. Doses of 5–10 μg of enterotoxin per animal were required to elicit emesis in 50 % of the monkeys tested (ED_{50}) using this model [90]. Bergdoll also noted that intravenous administration of purified SEs resulted in emesis at doses 10–100 times lower than that required for oral administration. The monkey feeding assay is covered in more detail in Chapter 11 of this volume. Emesis can also be elicited in monkeys via intravenous administration, but it does not provide a good model for foodborne illness [51].

Weanling piglets (0.9–9.1 kg in weight) have likewise been used to demonstrate emesis after dosing them with purified enterotoxin through curved feeding needles that reached the back of the throat [92]. In this model, the ED_{50} was 40–50 μg SEA per animal, which is higher than that required for monkeys, however the ready availability of pigs may make this a handier model system. Additionally, the authors mentioned that pigs may not become resistant to the enterotoxins as has been noted for monkeys, therefore it is possible that pigs can be used multiple times for different or repeat assays [90, 92]. As was seen with the TSS models in pigs, intravenous exposure to purified enterotoxins can also elicit emesis [83, 93].

Ferrets (~735 g) were used as an emesis model for SEB, in which doses of SEB were administered in a 10 ml volume into the stomach via an oral dosing tube [94]. The amount of enterotoxin required to elicit emesis in ferrets is approximately 5 mg/kg, which is higher than that required for primates or pigs. Lower doses (30 μg/kg) of enterotoxin have been shown to elicit emesis in ferrets when administered intravenously [95].

A more recent emesis model has been the house musk shrew (*Suncus murinus*, also known as shrew mice) [96–98]. Adult shrew mice (2–5 months old) that weigh 40–70 g are given purified enterotoxin in a 200 μl volume via a curved feeding needle into the back of the throat [96]. The ED_{50} of SEA was found to be 32 μg/kg and vomiting started just over an hour after administration of the toxin. Intraperitoneal administration of SEA lowered the ED_{50} to 3 μg/kg. Intraperitoneal administration of other SEs indicated that the house musk shrew displays a variable response to the enterotoxins based on which toxin is being administered, where SEA, SEE, and SEI required lower doses than did SEB, SEC2, SED, SEG, and SEH [98]. Of note, the doses used to elicit emesis in house musk shrews do not elicit signs of TSS or death.

4.4 Inflammatory Skin Disease Models

Ten-week-old BALB/c mice were used to study the response to SEB in order to examine the role of superantigens in the development of inflammatory skin diseases [99]. This study utilized intradermal injections and epicutaneous exposure of the mice to the superantigen and found that both administration methods elicited an inflammatory response in the skin, but that epicutaneous exposure required a tenfold higher amount (50–500 ng) of superantigen to elicit the same response as intradermal administration (5–50 ng) [99]. Both methods led to endocytic activation of Langerhan's cells, mast cell degranulation, and T-cell activation, and the local draining lymph nodes also became enlarged with an increased number of cells, indicating a strong inflammatory response to the superantigen. A modified version of this model was used by Savinko et al. and Laouini et al. in which BALB/c mice were exposed epicutaneously to SEB for a period of 1 week, the

SEB was removed for 2 weeks, and then the mice were resensitized for another week [100, 101]. This process was repeated for a total of 3 sensitization weeks, with 2-week breaks in between each exposure. SEB concentrations ranged from 0.5 to 10 μg of SEB, depending on the research group. Additionally, Savinko et al. looked at concurrent administration of the allergen ovalbumin (OVA) to see if superantigens changed the normal response to allergens in the skin. While Laouini et al. demonstrated a Th2-dominant cytokine response in the mice to SEB, Savinko et al. showed that SEB could elicit a mixed Th1/Th2 cytokine response; both groups identified an infiltration of T cells and induction of specific IgE (a hallmark of atopic dermatitis) in the skin using this model [100].

A slightly different model for studying inflammatory skin diseases was developed using human-severe combined immunodeficiency (SCID) mice reconstituted with human PBMCs from atopic dermatitis patients sensitized to the house dust mite allergen Der p [102]. The cells were injected intraperitoneally and intradermally in an attempt to reconstitute the skin mucosa as SCID mice do not have fully functioning lymphatic systems. Mice were then exposed epicutaneously to Der p, SEB, or a combination of the two. This model indicated that SEB may actually inhibit production of specific IgE, in contrast to the above models. SEB still induced infiltration of T cells and proinflammatory responses in the skin, alone and in the presence of Der p, whereas this did not occur in mice reconstituted with PBMCs from non-atopic dermatitis patients [102].

Another SCID mouse model utilizes human skin that is transplanted onto the mouse in order to further study psoriasis [103]. Skin transplants can be taken from lesions or non-lesion areas of psoriasis patients, and the transplants have been found to maintain their phenotypes post-transplantation. Using this model, researchers have been able to examine the role of superantigens and staphylococcal exfoliative toxins in psoriasis by intradermally injecting the toxins (2 μg) into the transplanted tissue. This, coupled with an additional transfer of PBMCs that have been stimulated ex vivo with the toxins (100 ng/ml for 48 h) led to the development of psoriasis in non-lesion transplants [43, 44, 104, 105]. It is interesting to note that transplants that came from healthy donors did not develop psoriasis after stimulation with superantigens, similar to what is seen with the human-SCID atopic dermatitis mouse model discussed above, indicating that other factors that are already present in the skin or PBMC transplants are also responsible for the development of these inflammatory skin disorders [43, 104].

5 Concluding Remarks

Superantigens have been implicated in multiple staphylococcal and streptococcal diseases, thus the development of animal models to adequately assess the role of superantigens in human disease is critical. Variable responses of animals to superantigens have made the development of good animal models more difficult, but steps like the creation of MHC class II transgenic mice, which alleviated the issue of the natural resistance of mice to superantigens, will continue to push the field forward. Gaining a better understanding of how superantigens cause or contribute to diseases will allow researchers to develop better treatment or prevention strategies.

References

1. Choi YW, Kotzin B, Herron L et al (1989) Interaction of *Staphylococcus aureus* toxin "superantigens" with human T cells. Proc Natl Acad Sci U S A 86:8941–8945. doi:10.1073/pnas.86.22.8941
2. White J, Herman A, Pullen AM et al (1989) The V beta-specific superantigen staphylococcal enterotoxin B: stimulation of mature T cells and clonal deletion in neonatal mice. Cell 56:27–35
3. Scholl PR, Diez A, Geha RS (1989) Staphylococcal enterotoxin B and toxic shock syndrome toxin-1 bind to distinct sites on HLA-DR and HLA-DQ molecules. J Immunol 143:2583–2588
4. Fraser JD (1989) High-affinity binding of staphylococcal enterotoxins A and B to HLA-DR. Nature 339:221–223. doi:10.1038/339221a0
5. Callahan JE, Herman A, Kappler JW, Marrack P (1990) Stimulation of B10.BR T cells with superantigenic staphylococcal toxins. J Immunol 144:2473–2479
6. Choi Y, Lafferty JA, Clements JR et al (1990) Selective expansion of T cells expressing V beta 2 in toxic shock syndrome. J Exp Med 172:981–984
7. Poindexter NJ, Schlievert PM (1985) Toxic-shock-syndrome toxin 1-induced proliferation of lymphocytes: comparison of the mitogenic response of human, murine, and rabbit lymphocytes. J Infect Dis 151:65–72
8. Lina G, Bohach GA, Nair SP et al (2004) Standard nomenclature for the superantigens expressed by *Staphylococcus*. J Infect Dis 189:2334–2336. doi:10.1086/420852
9. McCormick JK, Yarwood JM, Schlievert PM (2001) Toxic shock syndrome and bacterial superantigens: an update. Annu Rev Microbiol 55:77–104. doi:10.1146/annurev.micro.55.1.77
10. Brosnahan AJ, Mantz MJ, Squier CA et al (2009) Cytolysins augment superantigen penetration of stratified mucosa. J Immunol 182:2364–2373. doi:10.4049/jimmunol.0803283
11. Commons RJ, Smeesters PR, Proft T et al (2014) Streptococcal superantigens: categorization and clinical associations. Trends Mol Med 20:48–62. doi:10.1016/j.molmed.2013.10.004
12. Paillot R, Darby AC, Robinson C et al (2010) Identification of three novel superantigen-encoding genes in *Streptococcus equi* subsp. zooepidemicus, szeF, szeN, and szeP. Infect Immun 78:4817–4827. doi:10.1128/IAI.00751-10
13. Cole BC, Kartchner DR, Wells DJ (1990) Stimulation of mouse lymphocytes by a mitogen derived from *Mycoplasma arthritidis* (MAM). VIII. Selective activation of T cells expressing distinct V beta T cell receptors from various strains of mice by the "superantigen" MAM. J Immunol 144:425–431
14. Abe J, Takeda T, Watanabe Y et al (1993) Evidence for superantigen production by *Yersinia pseudotuberculosis*. J Immunol 151:4183–4188
15. Donadini R, Fields BA (2007) *Yersinia pseudotuberculosis* superantigens. Chem Immunol Allergy 93:77–91. doi:10.1159/0000100859
16. Stuart PM, Woodward JG (1992) *Yersinia enterocolitica* produces superantigenic activity. J Immunol 148:225–233

17. Fraser JD, Proft T (2008) The bacterial superantigen and superantigen-like proteins. Immunol Rev 225:226–243. doi:10.1111/j.1600-065X.2008.00681.x
18. Schlievert PM, Bohach GA (2007) Staphylococcal and streptococcal superantigens. In: Kotb M, Fraser JD (eds) Superantigens: molecular basis for their role human disease. ASM Press, Washington, DC, pp 21–36
19. Kim J, Urban RG, Strominger JL, Wiley DC (1994) Toxic shock syndrome toxin-1 complexed with a class II major histocompatibility molecule HLA-DR1. Science 266:1870–1874
20. Shupp JW, Jett M, Pontzer CH (2002) Identification of a transcytosis epitope on staphylococcal enterotoxins. Infect Immun 70:2178–2186
21. Peterson ML, Ault K, Kremer MJ et al (2005) The innate immune system is activated by stimulation of vaginal epithelial cells with *Staphylococcus aureus* and toxic shock syndrome toxin 1. Infect Immun 73:2164–2174. doi:10.1128/IAI.73.4.2164-2174.2005
22. Brosnahan AJ, Schlievert PM (2011) Gram-positive bacterial superantigen outside-in signaling causes toxic shock syndrome. FEBS J 278:4649–4667
23. Brosnahan AJ, Schaefers MM, Amundson WH et al (2008) Novel toxic shock syndrome toxin-1 amino acids required for biological activity. Biochemistry 47:12995–13003
24. Spaulding AR, Lin Y-C, Merriman JA et al (2012) Immunity to *Staphylococcus aureus* secreted proteins protects rabbits from serious illnesses. Vaccine 30:5099–5109. doi:10.1016/j.vaccine.2012.05.067
25. Kaempfer R, Arad G, Levy R et al (2013) CD28: direct and critical receptor for superantigen toxins. Toxins (Basel) 5:1531–1542. doi:10.3390/toxins5091531
26. Arad G, Levy R, Nasie I et al (2011) Binding of superantigen toxins into the CD28 homodimer interface is essential for induction of cytokine genes that mediate lethal shock. PLoS Biol 9, e1001149. doi:10.1371/journal.pbio.1001149
27. Shands KN, Schmid GP, Dan BB et al (1980) Toxic-shock syndrome in menstruating women: association with tampon use and *Staphylococcus aureus* and clinical features in 52 cases. N Engl J Med 303:1436–1442. doi:10.1056/NEJM198012183032502
28. Davis JP, Chesney PJ, Wand PJ, LaVenture M (1980) Toxic-shock syndrome: epidemiologic features, recurrence, risk factors, and prevention. N Engl J Med 303:1429–1435. doi:10.1056/NEJM198012183032501
29. Hill DR, Brunner ME, Schmitz DC et al (2005) In vivo assessment of human vaginal oxygen and carbon dioxide levels during and post menses. J Appl Physiol 99:1582–1591. doi:10.1152/japplphysiol.01422.2004
30. Yarwood JM, Schlievert PM (2000) Oxygen and carbon dioxide regulation of toxic shock syndrome toxin 1 production by *Staphylococcus aureus* MN8. J Clin Microbiol 38:1797–1803
31. Schlievert PM, Blomster DA (1983) Production of staphylococcal pyrogenic exotoxin type C: influence of physical and chemical factors. J Infect Dis 147:236–242
32. Bergdoll M (1984) Toxic shock syndrome toxin. Lancet 324:691. doi:10.1016/S0140-6736(84)91241-8
33. Bergdoll MS, Crass BA, Reiser RF et al (1981) A new staphylococcal enterotoxin, enterotoxin F, associated with toxic-shock-syndrome *Staphylococcus aureus* isolates. Lancet 1:1017–1021
34. Schlievert PM (1986) Staphylococcal enterotoxin B and toxic-shock syndrome toxin-1 are significantly associated with non-menstrual TSS. Lancet 1:1149–1150
35. Lalich IJ, Sam-Agudu NA (2010) Community-acquired methicillin-resistant *Staphylococcus aureus* necrotizing fasciitis in a healthy adolescent male. Minn Med 93:44–46
36. Kim H-J, Kim D-H, Ko D-H (2010) Coagulase-positive staphylococcal necrotizing fasciitis subsequent to shoulder sprain in a healthy woman. Clin Orthop Surg 2:256–259. doi:10.4055/cios.2010.2.4.256
37. Kravitz GR, Dries DJ, Peterson ML, Schlievert PM (2005) Purpura fulminans due to *Staphylococcus aureus*. Clin Infect Dis 40:941–947. doi:10.1086/428573
38. Stevens DL, Tanner MH, Winship J et al (1989) Severe group A streptococcal infections associated with a toxic shock-like syndrome and scarlet fever toxin A. N Engl J Med 321:1–7. doi:10.1056/NEJM198907063210101
39. Hennekinne J-A, De Buyser M-L, Dragacci S (2012) *Staphylococcus aureus* and its food poisoning toxins: characterization and outbreak investigation. FEMS Microbiol Rev 36:815–836. doi:10.1111/j.1574-6976.2011.00311.x
40. Kotler DP, Sandkovsky U, Schlievert PM, Sordillo EM (2007) Toxic shock-like syndrome associated with staphylococcal enterocolitis in an HIV-infected man. Clin Infect Dis 44:e121–e123. doi:10.1086/518286
41. Skov L, Olsen JV, Giorno R et al (2000) Application of Staphylococcal enterotoxin B on normal and atopic skin induces up-regulation

of T cells by a superantigen-mediated mechanism. J Allergy Clin Immunol 105:820–826. doi:10.1067/mai.2000.105524
42. Leung DY, Walsh P, Giorno R, Norris DA (1993) A potential role for superantigens in the pathogenesis of psoriasis. J Invest Dermatol 100:225–228
43. Boehncke WH (2001) Biologic effects of bacterial superantigens in a xenogeneic transplantation model for psoriasis. J Investig Dermatol Symp Proc 6:231–232. doi:10.1046/j.0022-202x.2001.00042.x
44. Boehncke WH, Hardt-Weinelt K, Nilsson H et al (2001) Antagonistic effects of the staphylococcal enterotoxin a mutant, SEA(F47A/D227A), on psoriasis in the SCID-hu xenogeneic transplantation model. J Invest Dermatol 116:596–601. doi:10.1046/j.1523-1747.2001.01295.x
45. Assimacopoulos AP, Strandberg KL, Rotschafer JH, Schlievert PM (2009) Extreme pyrexia and rapid death due to *Staphylococcus aureus* infection: analysis of 2 cases. Clin Infect Dis 48:612–614. doi:10.1086/597009
46. Kim SW, Grant JE, Kim SI et al (2004) A possible association of recurrent streptococcal infections and acute onset of obsessive-compulsive disorder. J Neuropsychiatry Clin Neurosci 16:252–260. doi:10.1176/appi.neuropsych.16.3.252
47. Vu BG, Stach CS, Kulhankova K et al (2015) Chronic superantigen exposure induces systemic inflammation, elevated bloodstream endotoxin, and abnormal glucose tolerance in rabbits: possible role in diabetes. mBio 6:1–10. doi:10.1128/mBio.02554-14
48. Vu BG, Stach CS, Salgado-Pabón W et al (2014) Superantigens of *Staphylococcus aureus* from patients with diabetic foot ulcers. J Infect Dis 210:1920–1927. doi:10.1093/infdis/jiu350
49. Vu BG, Gourronc FA, Bernlohr DA et al (2013) Staphylococcal superantigens stimulate immortalized human adipocytes to produce chemokines. PLoS One 8:e77988. doi:10.1371/journal.pone.0077988
50. De Azavedo JC (1989) Animal models for toxic shock syndrome: overview. Rev Infect Dis 11(Suppl 1):S205–S209
51. Quimby F, Nguyen HT (1985) Animal studies of toxic shock syndrome. Crit Rev Microbiol 12:1–44. doi:10.3109/10408418509104424
52. Beisel WR (1972) Pathophysiology of staphylococcal enterotoxin, type B, (SEB) toxemia after intravenous administration to monkeys. Toxicon 10:433–440
53. Pollack M, Weinberg WG, Hoskins WJ et al (1983) Toxinogenic vaginal infections due to *Staphylococcus aureus* in menstruating rhesus monkeys without toxic-shock syndrome. J Infect Dis 147:963–964
54. Mapara M, Thomas BS, Bhat KM (2012) Rabbit as an animal model for experimental research. Dent Res J (Isfahan) 9:111–118. doi:10.4103/1735-3327.92960
55. Kim YB, Watson DW (1970) A purified group A streptococcal pyrogenic exotoxin. Physiochemical and biological properties including the enhancement of susceptibility to endotoxin lethal shock. J Exp Med 131:611–622
56. Schlievert PM (1982) Enhancement of host susceptibility to lethal endotoxin shock by staphylococcal pyrogenic exotoxin type C. Infect Immun 36:123–128
57. Schlievert PM, Jablonski LM, Roggiani M et al (2000) Pyrogenic toxin superantigen site specificity in toxic shock syndrome and food poisoning in animals. Infect Immun 68:3630–3634
58. Arko RJ, Rasheed JK, Broome CV et al (1984) A rabbit model of toxic shock syndrome: clinicopathological features. J Infect 8:205–211
59. Scott DF, Kling JM, Kirkland JJ, Best GK (1983) Characterization of *Staphylococcus aureus* isolates from patients with toxic shock syndrome, using polyethylene infection chambers in rabbits. Infect Immun 39:383–387
60. Rasheed JK, Arko RJ, Feeley JC et al (1985) Acquired ability of *Staphylococcus aureus* to produce toxic shock-associated protein and resulting illness in a rabbit model. Infect Immun 47:598–604
61. Lee PK, Deringer JR, Kreiswirth BN et al (1991) Fluid replacement protection of rabbits challenged subcutaneous with toxic shock syndrome toxins. Infect Immun 59:879–884
62. Lee PK, Schlievert PM (1989) Quantification and toxicity of group A streptococcal pyrogenic exotoxins in an animal model of toxic shock syndrome-like illness. J Clin Microbiol 27:1890–1892
63. Parsonnet J, Gillis ZA, Richter AG, Pier GB (1987) A rabbit model of toxic shock syndrome that uses a constant, subcutaneous infusion of toxic shock syndrome toxin 1. Infect Immun 55:1070–1076
64. Sriskandan S, Unnikrishnan M, Krausz T et al (2001) Enhanced susceptibility to superantigen-

associated streptococcal sepsis in human leukocyte antigen-DQ transgenic mice. J Infect Dis 184:166–173. doi:10.1086/322018

65. Welcher BC, Carra JH, DaSilva L et al (2002) Lethal shock induced by streptococcal pyrogenic exotoxin A in mice transgenic for human leukocyte antigen-DQ8 and human CD4 receptors: implications for development of vaccines and therapeutics. J Infect Dis 186:501–510. doi:10.1086/341828
66. DaSilva L, Welcher BC, Ulrich RG et al (2002) Humanlike immune response of human leukocyte antigen-DR3 transgenic mice to staphylococcal enterotoxins: a novel model for superantigen vaccines. J Infect Dis 185:1754–1760. doi:10.1086/340828
67. Chen JY, Qiao Y, Komisar JL et al (1994) Increased susceptibility to staphylococcal enterotoxin B intoxication in mice primed with actinomycin D. Infect Immun 62: 4626–4631
68. Galanos C, Freudenberg MA, Reutter W (1979) Galactosamine-induced sensitization to the lethal effects of endotoxin. Proc Natl Acad Sci U S A 76:5939–5943
69. Lehmann V, Freudenberg MA, Galanos C (1987) Lethal toxicity of lipopolysaccharide and tumor necrosis factor in normal and D-galactosamine-treated mice. J Exp Med 165:657–663
70. Miethke T, Wahl C, Heeg K et al (1992) T cell-mediated lethal shock triggered in mice by the superantigen staphylococcal enterotoxin B: critical role of tumor necrosis factor. J Exp Med 175:91–98
71. Keppler D, Lesch R, Reutter W, Decker K (1968) Experimental hepatitis induced by D-galactosamine. Exp Mol Pathol 9:279–290
72. Record CO, Alberti KG, Williamson DH (1972) Metabolic studies in experimental liver disease resulting from D(+)-galactosamine administration. Biochem J 130:37–44
73. Tilahun AY, Marietta EV, Wu T-T et al (2011) Human leukocyte antigen class II transgenic mouse model unmasks the significant extrahepatic pathology in toxic shock syndrome. Am J Pathol 178:2760–2773. doi:10.1016/j.ajpath.2011.02.033
74. Huzella LM, Buckley MJ, Alves DA et al (2009) Central roles for IL-2 and MCP-1 following intranasal exposure to SEB: a new mouse model. Res Vet Sci 86:241–247. doi:10.1016/j.rvsc.2008.07.020
75. Krakauer T, Buckley M, Fisher D (2010) Murine models of staphylococcal enterotoxin B-induced toxic shock. Mil Med 175: 917–922
76. Mangalam AK, Rajagopalan G, Taneja V, David CS (2008) HLA class II transgenic mice mimic human inflammatory diseases. Adv Immunol 97:65–147. doi:10.1016/S0065-2776(08)00002-3
77. Nooh MM, El-Gengehi N, Kansal R et al (2007) HLA transgenic mice provide evidence for a direct and dominant role of HLA class II variation in modulating the severity of streptococcal sepsis. J Immunol 178: 3076–3083
78. Kotb M, Norrby-Teglund A, McGeer A et al (2002) An immunogenetic and molecular basis for differences in outcomes of invasive group A streptococcal infections. Nat Med 8:1398–1404. doi:10.1038/nm800
79. Kasper KJ, Zeppa JJ, Wakabayashi AT et al (2014) Bacterial superantigens promote acute nasopharyngeal infection by *Streptococcus pyogenes* in a human MHC Class II-dependent manner. PLoS Pathog 10:e1004155. doi:10.1371/journal.ppat.1004155
80. Rajagopalan G, Smart MK, Murali N et al (2007) Acute systemic immune activation following vaginal exposure to staphylococcal enterotoxin B—implications for menstrual shock. J Reprod Immunol 73:51–59. doi:10.1016/j.jri.2006.06.007
81. Reipert BM, Steinitz KN, van Helden PM et al (2009) Opportunities and limitations of mouse models humanized for HLA class II antigens. J Thromb Haemost 7(Suppl 1):92–97. doi:10.1111/j.1538-7836.2009.03403.x
82. Bulanda M, Zaleska M, Mandel L et al (1989) Toxicity of staphylococcal toxic shock syndrome toxin 1 for germ-free and conventional piglets. Rev Infect Dis 11(Suppl 1):S248–S253
83. Van Gessel YA, Mani S, Bi S et al (2004) Functional piglet model for the clinical syndrome and postmortem findings induced by staphylococcal enterotoxin B. Exp Biol Med (Maywood, NJ) 229:1061–1071
84. Hudson LC, Seabolt BS, Odle J et al (2012) Sublethal staphylococcal enterotoxin B challenge model in pigs to evaluate protection following immunization with a soybean-derived vaccine. Clin Vaccine Immunol 20:24–32. doi:10.1128/CVI.00526-12
85. Salgado-Pabón W, Breshears L, Spaulding AR et al (2013) Superantigens are critical for *Staphylococcus aureus* infective endocarditis, sepsis, and acute kidney injury. mBio 4:e00494–13. doi:10.1128/mBio.00494-13

86. Strandberg KL, Rotschafer JH, Vetter SM et al (2010) Staphylococcal superantigens cause lethal pulmonary disease in rabbits. J Infect Dis 202:1690–1697. doi:10.1086/657156
87. Karau MJ, Tilahun AY, Schmidt SM et al (2012) Linezolid is superior to vancomycin in experimental pneumonia caused by Superantigen-Producing *Staphylococcus aureus* in HLA class II transgenic mice. Antimicrob Agents Chemother 56:5401–5405. doi:10.1128/AAC.01080-12
88. Spaulding AR, Salgado-Pabón W, Merriman JA et al (2014) Vaccination against *Staphylococcus aureus* pneumonia. J Infect Dis 209:1955–1962. doi:10.1093/infdis/jit823
89. Fowler VG, Miro JM, Hoen B et al (2005) *Staphylococcus aureus* endocarditis: a consequence of medical progress. JAMA 293:3012–3021. doi:10.1001/jama.293.24.3012
90. Bergdoll MS (1988) Monkey feeding test for staphylococcal enterotoxin. Methods Enzymol 165:324–333
91. SURGALLA MJ, BERGDOLL MS, DACK GM (1953) Some observations on the assay of staphylococcal enterotoxin by the monkey-feeding test. J Lab Clin Med 41:782–788
92. Taylor SL, Schlunz LR, Beery JT et al (1982) Emetic action of staphylococcal enterotoxin A on weanling pigs. Infect Immun 36:1263–1266
93. Kohrman KA, Kirkland JJ, Danneman PJ (1989) Response of various animal species to experimental infection with different strains of *Staphylococcus aureus*. Rev Infect Dis 11(Suppl 1):S231–S236, discussion S236–7
94. Wright A, Andrews PL, Titball RW (2000) Induction of emetic, pyrexic, and behavioral effects of *Staphylococcus aureus* enterotoxin B in the ferret. Infect Immun 68:2386–2389
95. Sagrada A, Rizzi C, Bonali P, Donetti A (1995) 5-HT3 receptor antagonists inhibit enterotoxin B-induced emesis in the ferret. In: Reynolds DJM, Andrews PLR, Davis CJ (eds) Serotonin and the scientific basis of anti-emetic therapy. Oxford Clinical Communications, Oxford, pp 256–257
96. Hu DL, Omoe K, Shimura H et al (1999) Emesis in the shrew mouse (*Suncus murinus*) induced by peroral and intraperitoneal administration of staphylococcal enterotoxin A. J Food Prot 62:1350–1353
97. Hu D-L, Omoe K, Sashinami H et al (2009) Immunization with a nontoxic mutant of staphylococcal enterotoxin A, SEAD227A, protects against enterotoxin-induced emesis in house musk shrews. J Infect Dis 199:302–310. doi:10.1086/596065
98. Hu D-L, Omoe K, Shimoda Y et al (2003) Induction of emetic response to staphylococcal enterotoxins in the house musk shrew (*Suncus murinus*). Infect Immun 71:567–570
99. Saloga J, Leung DY, Reardon C et al (1996) Cutaneous exposure to the superantigen staphylococcal enterotoxin B elicits a T-cell-dependent inflammatory response. J Invest Dermatol 106:982–988
100. Savinko T, Lauerma A, Lehtimäki S et al (2005) Topical superantigen exposure induces epidermal accumulation of CD8+ T cells, a mixed Th1/Th2-type dermatitis and vigorous production of IgE antibodies in the murine model of atopic dermatitis. J Immunol 175:8320–8326
101. Laouini D, Kawamoto S, Yalcindag A et al (2003) Epicutaneous sensitization with superantigen induces allergic skin inflammation. J Allergy Clin Immunol 112:981–987. doi:10.1016/j.jaci.2003.07.007
102. Herz U, Schnoy N, Borelli S et al (1998) A human-SCID mouse model for allergic immune response bacterial superantigen enhances skin inflammation and suppresses IgE production. J Invest Dermatol 110:224–231. doi:10.1046/j.1523-1747.1998.00119.x
103. Nickoloff BJ, Kunkel SL, Burdick M, Strieter RM (1995) Severe combined immunodeficiency mouse and human psoriatic skin chimeras. Validation of a new animal model. Am J Pathol 146:580–588
104. Boehncke WH, Dressel D, Zollner TM, Kaufmann R (1996) Pulling the trigger on psoriasis. Nature 379:777. doi:10.1038/379777a0
105. Wrone-Smith T, Nickoloff BJ (1996) Dermal injection of immunocytes induces psoriasis. J Clin Invest 98:1878–1887. doi:10.1172/JCI118989

Chapter 2

Identification, Purification, and Characterization of Staphylococcal Superantigens

Joseph A. Merriman and Patrick M. Schlievert

Abstract

Purifying natively produced staphylococcal superantigens is an important process in the study of these proteins, as many common methods of protein purification are affected by staphylococcal protein A contamination. Here, we describe a proven approach for identifying superantigens in vitro as well as for purifying novel superantigens both in His-tagged and native forms using modern genetic tools coupled with thin-layer isoelectric focusing.

Key words Staphylococcal superantigen, Purification, Identification, Characterization, Polyclonal antibody production

1 Introduction

Superantigens function as potent stimulators of the immune system and significantly contribute to disease progression in food poisoning [1], toxic shock syndrome [2], endocarditis [3], and pneumonia [4]. A new staphylococcal superantigen, staphylococcal enterotoxin-like X, has been described as recently as 2011 and was shown to contribute significantly to necrotizing pneumonia in the endemic methicillin-resistant *Staphylococcus aureus* (MRSA) strain USA300 [4]. This recent discovery, in conjunction with ~50 % of the *S. aureus* secretome being undefined [5], warrants a standard method in identifying, purifying, and characterizing new staphylococcal superantigens as they arise, as well as already known ones that are needed for pathogenesis studies.

Superantigens range in size from 19 to 30 kDa and lack post-translational modification. Their molecular weights can make them inherently difficult to identify by Western blot in a complex mixture of proteins due to the presence of protein A. Protein A is a staphylococcal protein that is both cell surface-associated and -secreted, having molecular weights of 15, 30, 45, and 60 kDa. Protein A is capable of binding the Fc portion of antibodies,

Amanda J. Brosnahan (ed.), *Superantigens: Methods and Protocols*, Methods in Molecular Biology, vol. 1396, DOI 10.1007/978-1-4939-3344-0_2,

making Western blot analysis difficult when desired superantigens fall in the same molecular weight as protein A. Our method described herein is capable of separating superantigens from the contaminating protein A based on their neutral to basic isoelectric properties.

Superantigen purification also benefits from the molecules' stabilities under many denaturing conditions such as boiling, pepsin/trypsin treatment, desiccation, and ethanol precipitation. Our protocol is able to harness the stability of superantigens during ethanol treatment to remove many small and large molecule contaminants. This precipitate can then be solubilized in water and proteins separated based on pH gradients set up by ampholytes in isoelectric focusing under continuous current. The purity of this natively expressed protein is typically ≥99 %, as determined by Coomassie blue R250 and silver staining.

As new superantigens arise, it is imperative to have a standard method to identify their production in culture fluids through stimulation of peripheral blood mononuclear cells (PBMCs) from humans, fever production in rabbits, and endotoxin enhancement in rabbits. It has been previously demonstrated that administering lipopolysaccharide (LPS), approximately 2–4 h after administration of a superantigen, synergizes to increase lethality by up to 10^6-fold [6]. These approaches have inherent difficulties with detecting superantigenicity because *S. aureus* is also capable of producing potent cytotoxins. Our protocols presented here have ways to avoid inadvertent proinflammatory responses from cytotoxins, allowing successful identification of superantigen-containing culture fluids.

Generation of antibodies to desired superantigens is important once purified proteins are isolated. IEF, as defined in this methodology chapter, or through use of His-tagged recombinant technology, as discussed in a previously published methodology [7], are both suitable sources of proteins for generating superantigen-specific polyclonal antibodies from rabbits. Previous work by our laboratory has demonstrated strong antibody responses develop to superantigens in sublethal-dose vaccination studies [8, 9], thus allowing us to use the same protein amounts to generate specific polyclonal antibodies, which in turn can be subsequently used for many immunological techniques.

2 Materials

2.1 Identification of New Staphylococcal Superantigens

All glassware and reagents *must be* maintained pyrogen-free to avoid LPS contamination.

1. 25 mL overnight culture grown in Todd Hewitt broth at 37 °C with aeration, of suspected superantigen-producing *S. aureus* strain.

2. 25 mL defibrinated rabbit red blood cells.
3. 0.45 μm filters.
4. 0.8 % agarose in phosphate-buffered saline (PBS) kept at 50 °C.
5. 50 mL polypropylene conical tubes.
6. Glass microscope slides.
7. 100 mm × 15 mm Petri dish.
8. 5 mL serological pipette and dispenser.
9. 4 mm vacuum-assisted hole punch.
10. 6 New Zealand white rabbits (2 kg each).
11. 1 mL syringes.
12. 25-gauge needles.
13. Lipopolysaccharide (LPS): isolated from *Salmonella enterica* serovar Typhimurium, 1 mg/mL solubilized in PBS.
14. Fast reading, electronic rectal thermometer.

2.2 Superantigen Purification

All materials *must be* maintained pyrogen-free to avoid LPS contamination.

1. *Dialyzed Beef Heart (BH) growth medium:* Todd-Hewitt medium is useful in everyday studies of staphylococcal cultures however, for the purpose of superantigen purification it is optimal to use a dialyzed medium. Many medium proteins fall within a molecular weight of 20–50 kDa and superantigens are also within this range. Excessive medium protein can contaminate and cause false readings when determining protein purity by Coomassie blue R250 or silver stain.

   To bypass this problem we use a dialyzed beef heart (BH) medium, made directly from bovine hearts which is summarized below [10].

   (a) Add 10 lb of ground beef heart to 3 L of pyrogen-free H_2O and heat to 70 °C.

   (b) Stir tissue constantly, to avoid burning, for 30 min at 70 °C.

   (c) Cool the mixture to ~45 °C.

   (d) Add 16 g of porcine pancreas trypsin to proteolyze the proteins in the tissue.

   (e) A pH of 8 must be maintained via titration of 10 M NaOH, as needed, with vigorous stirring for 2 h.

   (f) After 2 h, bring the pH to 8.5, cover with foil and store at 4 °C for 18 h.

   (g) The next day, remove mixture from 4 °C and remove any large masses of fat floating on the top as this can affect the dialysis process.

(h) Be sure to adjust the pH to 8.0 if the pH has dropped below that point.

(i) Slowly add the trypsinized tissue mixture to 3 ft long segments of 45 mm, 12–14 k molecular weight (MW) cut-off dialysis tubing (*see* **Note 1**).

(j) After filling tubes, place in baked 20 L glass carboys with an equal number of dialysis tubing per carboy.

(k) Fill each carboy with 10 L of pyrogen-free water and dialyze for 4–5 days at 4 °C.

(l) After dialysis, the dialysate should be brown. This will contain small molecular weight BH proteins and other essential factors; the dialysate is used as the preferred microbiological medium.

(m) Combine the dialysates from each carboy and mix well.

(n) We find 1200 mL of BH media in 3 L Fehrenbach flasks provides the optimal growth conditions for aeration and superantigen production in ambient air. This volume will be used throughout the protocol.

2. All glassware is heated (baked) to 165 °C for a minimum of 3 h before use to avoid LPS contamination. Any items that cannot be heated are soaked in 10 % sodium dodecyl sulfate solution for 30 min and washed with pyrogen-free water.

3. Two to four 3 L Fehrenbach flasks containing 1200 ml of BH medium each (as prepared in Subheading 2.2, **item 1**).

   (a) BH medium is supplemented 1:100 (v/v; 1 % final volume) with glucose-phosphate buffer.

   (b) Glucose-phosphate buffer: 0.33 M dextrose, 0.5 M $NaHCO_3$, 0.68 M NaCl, 0.12 M $Na_2HPO_4{\cdot}7H_2O$, 0.027 M L-Glutamine.

4. Overnight culture of bacterial isolate grown in 25 ml Todd Hewitt broth.

5. Approximately 20 L of 200 proof ethanol for toxin precipitation.

6. Washed G-75 Sephadex: prepared by following steps. Washing is done to completely dehydrate and collapse the Sephadex beads. The collapsed Sephadex is used as a gel matrix for proteins to migrate through without being hindered by molecular weight.

   (a) Swell Sephadex in a 4 L flask with 4 L of water overnight at 4 °C.

   (b) Exhaustively dehydrate with ethanol the next day using a vacuum collection flask with Whatman paper in a funnel as a filter.

   (c) Allow Sephadex to dry to a fine powder in a chemical safety hood before straining to remove large aggregates and returning to original container.

7. Autoclaved, nano-purified deionized pyrogen-free water contained in baked glassware.
8. Spectra/Por dialysis membrane, 45 mm, 12–14 k MW cutoff and 23 mm, 6–8 k MW cutoff (Spectrum Laboratories Inc., Rancho Dominguez, CA)
9. Pharmacia-LKB Biotechnology Multiphor 2117 thin-layer isoelectric focusing apparatus able to accommodate a plate size of 26 × 12.5 cm (Amersham Pharmacia Biotech, Piscataway, NJ). Plate needs to be maintained on a glass continuous water flow chamber at 2 °C. A constant power source capable of maintaining 1000 V, 20 mAmps, and 8 W for 24 h is required to run this machine.
10. A 26 × 12.5 cm glass bottom thin-layer isoelectric focusing plate with rubber boarders.
11. Isoelectric focusing (IEF) electrodes (GE Healthcare, Uppsala, Sweden).
12. IEF electrode strips.
13. Solutions of 50 mL 1 M NaOH and 50 mL of 1 M H_3PO_4 to set up the IEF electrodes.
14. Ampholytes with pH ranges 3–10, 6–8, and 7–9 (Sigma-Aldrich, St. Louis, MO).
15. 3MM Whatman paper.
16. 10 % trichloroacetic acid.
17. Coomasie R250 protein stain and destain.
18. Antibodies against your superantigen for Western blot analysis (if available).
19. Pyrex™ glass wool.
20. 0.45 μm syringe filter.

2.3 Mitogenicity Assay

Use only plastic materials when working with lymphocytes.

1. 50 mL human blood.
2. 50 mL polypropylene conical tubes.
3. 15 mL polypropylene tubes.
4. 200 proof ethanol.
5. Pyrogen-free H_2O.
6. 100 mL Histopaque-1077® solution (Sigma Aldrich, St. Louis, MO).
7. 500 mL Roswell Park Memorial Institute medium (RPMI)-1640.
8. 20 mL Complete RPMI: supplemented with 1 % Penicillin-Streptomycin (10,000 U/mL), 2 % Fetal Calf Serum, 20 μM glutamine.
9. Hemocytometer.

10. 96-well tissue culture plate.
11. 100 scintillation vials.
12. 300 mL Cytoscint™-ESLiquid Scintillation Cocktail (MP Biomedicals, LLC, Solon, OH).
13. Titertek® Plus Cell Harvester.
14. Pressed fiber-glass paper.
15. Scintillation counter capable of making counts per minute (cpm) readings.

2.4 Generating Polyclonal Antibodies in Rabbits

1. New Zealand White rabbit (under 6 months of age).
2. Freund's incomplete adjuvant.
3. Three aliquots 10–20 μg of superantigen in PBS at final volumes of 500 μl.
4. 3 ml sterile syringe.
5. 5 ml sterile glass culture tubes.
6. 18- and 20-gauge syringe needles.
7. 1 and 60 mL sterile syringes.
8. 50 ml conical centrifuge tube.
9. Sterile wooden applicator sticks.
10. Sterile microfuge tube.
11. Glass microscope slide.
12. 5 mL serological pipette and dispenser.
13. 4 mm vacuum-assisted hole punch.
14. 0.8 % agarose in PBS (kept at 50 °C until dispensed).
15. Non-specific protein of choice to test antibody specificity.
16. Ketamine 100 mg/ml.
17. Xylazine 100 mg/ml.
18. Ethasol® (Active ingredients 390 mg/ml pentobarbital sodium and 50 mg/ml phentoin sodium) (Vibrac AH, Inc., Fort Worth, Texas).

3 Methods

3.1 Identification of New Staphylococcal Superantigens

1. Centrifuge overnight culture of *S. aureus* strain (25 ml) at 4000 × *g* for 10 min.
2. Remove bacterial supernate and filter-sterilize through a 0.45 μm filter.
3. Add 1:1 defibrinated rabbit red blood cells to bacterial supernate and rock gently at 37 °C for 1 h.
4. Centrifuge blood/supernate mixture at 4000 × *g* for 10 min.
5. Remove supernate and filter-sterilize through a 0.45 μm filter.

6. Using 0.8 % agarose in PBS, add rabbit red blood cells to a final concentration of 3 % red blood cells.
7. Quickly dispense 4 ml of agarose/blood mixture onto microscope slides.
8. Allow agarose/blood mixture to solidify at room temperature. Keep solidified agarose slide in 100 mm × 15 mm Petri dish on top of a moist paper towel to provide a humidified environment.
9. Using a vacuum-assisted hole punch, punch a 4 mm hole into the agarose, remove the plug, and add 20 μl of supernate from **step 5**.
10. Incubate at 37 °C for 4 h.
11. If major cytolysins have been removed, no zone of clearing should be seen surrounding the supernate.
12. If zones of clearing are present, repeat **steps 1–5** as needed to remove residual cytolysins.
13. Once cytolysins have been removed, make 1:10 and 1:100 serial dilutions from the original supernate.
14. Inject, intravenously, two rabbits per group with 2 mL of each dilution (undiluted, 1/10, and 1/100), as well as two PBS-only controls (*see* **Note 2**).
15. Measure fever responses hourly over 4 h, noting that superantigens typically cause linear rises in body temperatures, peaking at 4 h (*see* **Note 3**).
16. After 4 h, intravenously inject each rabbit, excluding PBS control, with 5 μg of LPS (1/100 of LD_{50} dose) (*see* **Note 4**).
    (a) LPS is given at this time point because LPS and superantigen synergism is optimal at this point and will cause rapid lethality if superantigen is present.
17. Assess lethality over a 48-h period (*see* **Note 5**).
18. If fever and enhancement of LPS lethality are seen, proceed to purification steps.

3.2 Purification of Staphylococcal Superantigens

1. Inoculate 2–4 × 1200 mL flasks of BH plus 1 % glucose-phosphate buffer with overnight culture of desired superantigen-producing *S. aureus*.
2. Incubate cultures for 48 h at 37 °C at 220 × g (*see* **Note 6**).
3. After 48 h incubation, pour cultures into 6 L Erlenmeyer flasks already containing 4800 mL of 100 % ethanol, and let cultures precipitate for 2–5 days at room temperature or 4 °C, until liquid in flasks is transparent and mats of precipitate are at the bottoms (*see* **Note 7**).
4. Slowly pour off and discard the majority of clear supernate without agitating the precipitate into solution (*see* **Note 8**).

5. Collect precipitate in 500 mL plastic centrifuge bottles and centrifuge for 10 min at 4000 × *g* at 4 °C (*see* **Note 9**).
6. Allow pellets to air-dry for 2–3 h to remove ethanol. Pellet is considered free of ethanol when the pellet surface looks matte but not completely dry.
7. Add 75 mL of pyrogen-free H_2O and stir pellet with a magnetic stir bar and plate at 1000 × *g* for a minimum of 4 h at room temperature.
8. Remove stir bar and centrifuge at 8000 × *g* for 30 min at 4 °C; save the supernate and discard pellet.
9. Dialyze supernate in 45 mm, 12,000–14,000 MW cutoff dialysis tubing in 4 L pyrogen-free water overnight at 4 °C with gentle stirring (*see* **Note 1**).
10. Dry dialyzed sample (contents of the dialysis bag) to ~75–90 mL (*see* **Note 10**).
11. Empty sample into a 150 mL beaker and add 2.5 mL of pH 3–10 ampholytes.
12. Add approximately 4.5 g washed G-75 Sephadex to sample, stirring slowly to avoid air bubbles (*see* **Note 11**).
13. Cut 3 IEF electrode strips for each end of IEF plate, each a little smaller than the plate to allow for expansion after wetting.
14. Wet electrode strips with ~2.5 ml pyrogen-free water.
15. Pour Sephadex-containing sample into IEF plate and spread/smooth with a flat-edged spatula and light tapping plate on bench top to make a completely smooth surface.
16. Dry plate until the surface no longer looks glossy (should have a matte appearance). (If cracks occur *see* **Note 12.**)
17. After drying, cut two more wicks (slightly longer than the prior three wicks and slightly shorter than plate) for dipping in either 1 M NaOH or 1 M H_3PO_4 for placement on ends of the IEF plate; these will be the anionic and cationic ends, respectively.
18. Place IEF plate on thin-layer IEF water coolant plate.
19. Set the maximum settings: Volts → 1000; Watts → 8; mAmps → 20.
20. Let IEF run for 24 h.
21. Remove plate from apparatus, and with a flat-edged spatula, scrape non-cloudy portion of gel (NaOH end) into a new 150 ml beaker (*see* **Note 13**).
22. Bring non-cloudy fraction to 100 mL using sterile pyrogen-free H_2O and repeat **steps 11–20**, except this time use ampholytes with a pH range of 6–8 or 7–9, depending on isoelectric point of the superantigen you are trying to purify.

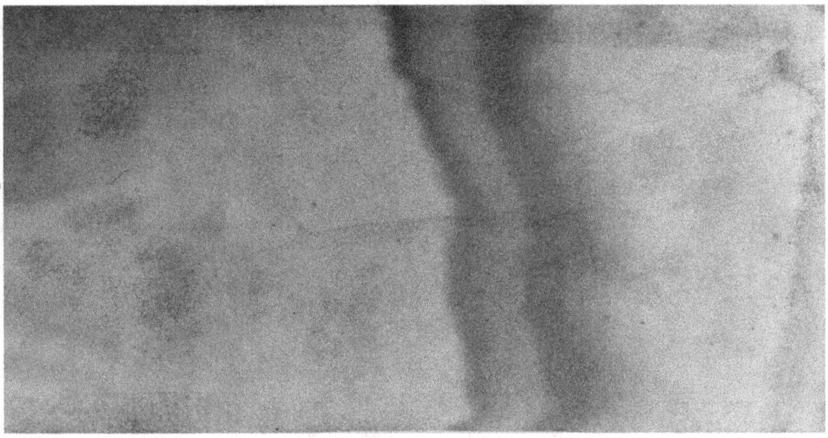

Fig. 1 Zymogram print of isoelectric focusing of TSST-1 during purification process. This is the secondary separation at pH 6–8. The two bands are TSST-1 with a difference in pH of 0.04, showing the high resolving power of this technique

23. Typically, after 18–24 h, you will notice a clear band in the second gel due to the light diffraction properties caused by the focusing of the superantigen. This band can be carefully scraped into a 50 mL conical tube, suspended in 10 mL pyrogen-free water, and then you can skip to **step 26**. If not proceed to next step.
24. The highest concentration of protein can be detected at this point through the use of a zymogram print on 3 MM Whatman paper. Briefly, over-lay plate with Whatman paper, wash for ~30 s with 10 % trichloroacetic acid and stain with Coomassie R250 and destain yielding a pattern seen in Fig. 1. This was a secondary separation at pH 6–8. The two bands are TSST-1 with a difference in pH of 0.04, showing the high resolving power of this technique.
25. Grid the plate into roughly 15 equal-sized fractions (from anode to cathode), and using a flat-edged spatula, scrape each into 50 mL conical tubes, rinsing spatula between fractions.
26. Suspend each fraction in 5 mL of sterile pyrogen-free H_2O, invert 5–10 times, and let settle at 4 °C overnight.
27. Carefully pipette ~10 μl of supernate out of each tube, avoiding Sephadex (*see* **Notes 14** and **15**), and run on SDS-PAGE to examine protein content. At the same time, the supernate can be placed in a pH electrode for isoelectric point determination.
28. Identify fractions of desired protein in highest concentrations through Coomassie blue R250 and Western blot (if antibodies for the superantigen you are purifying are available) and then pool those fractions.

29. Pack 1 cm of glass wool into the bottom of a 60 mL syringe. Tape needle-tip end down to the top of a 50 mL conical tube and pour desired fractions in to allow gel filtration to occur. When most of the liquid has run through, plunge Sephadex/protein mix through the glass wool to remove the majority of the remaining liquid. This step will remove most of the Sephadex from the purified superantigen (superantigen remains in liquid portion).
30. Filter liquid containing your superantigen through a 0.45 μm pore size syringe filter.
31. Dialyze the purified superantigen in 6–8 k MW cutoff tubing for 4 days at 4 °C in 1 L pyrogen-free water with slow continuous stirring (*see* **Note 16**).
32. Quantify desired protein using Bradford reagent with a standard curve generated using a previously described superantigen for best accuracy.
33. An additional SDS-PAGE gel with subsequent Coomassie blue R250 staining should reveal a protein purity of ≥99 %.

3.3 Lymphocyte Mitogenicity Assay

The following procedure should be carried out in biological safety cabinet.

1. Draw 50 mL of human blood into 5 mL heparin to prevent clotting.
2. Dilute blood 1:1 with RPMI-1640 medium at room temperature into two 50 mL conical centrifuge tubes and mix gently by inversion.
3. At room temperature, pour 18 mL of Histopaque® into 4 × 50 mL tubes and carefully layer 24 mL of diluted blood on top (*see* **Note 17**).
4. Centrifuge at 400 × *g* for 30 min
5. Collect 5–6 mL of the buffy coat (central interface) by pipetting slowly, and combine in 50 mL centrifuge tube. Discard the remaining blood components.
6. Dilute buffy coat with RPMI-1640 to 50 mL and centrifuge at 400 × *g* for 10 min.
7. Discard supernate.
8. Resuspend pellet (lymphocytes) by gently pipetting 10 mL RPMI-1640 into the tube and gently pipetting up and down against the pellet.
9. Pipette resuspended cells into a 15 mL centrifuge tube and add RPMI-1640 to 15 mL total.
10. Centrifuge at 400 × *g* for 10 min.
11. Decant supernate and resuspend pellet in 15 mL RPMI-1640.

12. Centrifuge at 400 × *g* for 10 min.
13. Decant supernate and resuspend pellet in 5 mL of *complete* RPMI.
14. Count the lymphocytes in a hemocytometer; adjust cells to a concentration of 2×10^5 cells/200 μL.
15. Add 200 μL of lymphocytes to 96-well, flat-bottom microtiter plates (quantity of wells will depend on total tests being carried out).
16. Serial dilute superantigen-containing fractions tenfold (final dilution 10^{-5}) from purification process to be added in 20 μl volumes per well of the microtiter plates (test samples in triplicate or quadruplicate) (*see* **Note 18**).
17. Serial dilute 1 mg/ml staphylococcal enterotoxin B (SEB) as a positive control (you want 1 μg/well–10^{-6} μg/well) and diluent from fractions (usually pyrogen-free water) as a negative control.
18. Each well will contain the following:
    - 20 μl of toxin dilution or control to appropriate well
    - 200 μl of diluted lymphocytes to each well
19. Incubate in 5 % CO_2 at 37 °C for 3 days.
20. On day 3, add 1 μCi tritiated thymidine to each well in a 20 μl volume.
21. Incubate in 5 % CO_2 at 37 °C for 24 h.
22. Titertek® Plus Cell Harvester is used to harvest lymphocyte population. First load empty scintillation vials into a vial holder to correspond with plate design.
23. Turn on vacuum attached to cell harvester, then turn the harvester on.
24. Fill two metal dishes, one with distilled H_2O and one with absolute ethanol.
25. Place white filter paper (fiberglass paper) on machine in designated places, and clamp it down to hold in place.
26. Use the vacuum to draw (~24 psi) H_2O onto the filter paper to wet it.
27. Use the vacuum to draw lymphocytes out of corresponding wells onto the filter papers, and wash with H_2O for seven fills of the wells.
28. Draw up ethanol onto the filter papers and pull vacuum for 1–2 min to help dry the filter paper.
29. Switch harvester to pull filter disks to the apparatus.
30. Turn off harvest switch, raise cut filter pieces, and place filter disks in corresponding tubes.

31. Repeat process for all wells
32. Let filter paper disks dry in tubes for at least 1 h
33. Add 3 mL of Cytoscint™-ESLiquid to each vial and cap.
34. Count on scintillation counter in cpm (or dmp) units (*see* **Note 19**).

3.4 Generating Superantigen-Specific Polyclonal Antibodies

1. Take ~1 mL of blood from a rabbit's marginal ear vein, to be used as a non-immune serum control.
2. Dispense 0.5 mL of Freund's incomplete adjuvant into glass culture tube.
3. Using 3 mL syringe and 18-gauge needle, add 0.5 mL of superantigen preparation to the incomplete adjuvant, and continuously pull up and push down (*see* **Note 20**).
4. Once mixture has become opaque and highly viscous, inject mixture in bolus doses, subcutaneously, into the nape of the neck of the rabbit (multiple sites).
5. **Steps 1–3** are repeated two more times on days 7 and 14 post primary immunization, for a total of three immunizations.
6. Seven days after final immunization, the rabbit is anesthetized using 0.8 ml Ketamine (100 mg/ml) and 0.4 ml Xylazine (100 mg/ml).
7. Direct cardiac puncture through the ribcage is performed using a 60 ml syringe and 18-gauge needle.
8. Exsanguinate the animal through this cardiac puncture. Roughly 80–100 mL of blood can be obtained from a 4 to 5 kg rabbit.
9. Our laboratory uses a 1 mL dose of Euthasol® (390 mg pentobarbital sodium and 50 mg phenytoin sodium) for premature euthanasia and then subsequent opening of the thoracic cavity to ensure animal expiration.
10. Blood should be distributed evenly across 50 ml conical tubes and stored at room temperature.
11. After 1 h at room temperature, ring around the blood clot using a sterile wooden applicator.
12. Tubes should then be stored at 4 °C for 18 h.
13. After 18 h, centrifuge tubes at 4000 × *g* for 30 min.
14. ***Carefully remove serum using a serological pipette and transfer to another conical tube.
15. Filter-sterilize serum using a 0.45 μm syringe filter.
16. Set up a double immunodiffusion assay by pipetting 4 mL molten 0.8 % agarose in PBS onto a microscope slide and allow it to harden at room temperature.
17. Carefully punch three 4 mm holes, 4 mm apart, into the agarose.

18. Add 20 μl of immunized serum to one end well and 20 μl of non-immunized serum on the other side.
19. In the center well, add 20 μl of a 100 μg/ml preparation of the purified toxin used for immunization.
20. Incubate slide in humidified chamber (Petri dish with moistened paper towel at the bottom) at 37 °C for 4-8 h.
21. Immunized serum should react with purified protein within 4-8 h, as depicted by a white immunoprecipitate between the serum and protein well.
22. Serum containing polyclonal antibody against superantigen should be aliquotted in 1 ml amounts and stored at -20 °C until needed.

4 Notes

1. All dialysis tubing should be thoroughly hydrated in water before adding any BH or protein sample. This can be done by submerging membranes in a beaker of water for ~1 min. Open the tubing by rubbing between your gloved fingers. Knot one end and fill with water to ensure no leaks are in the tubing before adding your sample.
2. LPS-only control rabbits can also be used.
3. In contrast, if there is contaminating LPS, the fever responses will peak first at 1 h and then at 3 h post-injection.
4. Alternatively, at the 4 h time point, the PBS rabbits may be injected with LPS to verify lack of enhanced lethality.
5. Synergistic effects of superantigen and LPS are potent. If a superantigen is present, it is likely all rabbits will die or be severely ill by 2-4 h, but we keep the animals for 48 h.
6. Secreted virulence factors, i.e. superantigens, are typically made in early stationary phase. Therefore, 48 h ensures they are produced to maximum amount and maximally released from the cell wall.
7. A 1 (culture):4 (ethanol) dilution is critical for ethanol precipitation to ensure small molecules <10 kDa remain in the supernate, while heavier molecules precipitate. This removes the majority of the medium proteins.
8. Precipitate will slide and collect on one side of the flask. This is expected but caution should be taken to avoid pouring out too much of the precipitate as this contains the desired superantigen.
9. This step will have to be repeated multiple times until all flasks are empty. Simply pour off supernate after each centrifugation, and add more precipitate.

10. Drying of dialysis tubing can be significantly expedited if done on a vent in a biological safety cabinet. Rotate tube frequently to avoid uneven drying and cracking.
11. Sample plus Sephadex should be cake batter consistency to allow easy spreading and smoothing of the surface.
12. Again, drying process can be expedited by drying on a vent in biological safety cabinet. If cracks occur, scrape out sample and rehydrate to prior consistency before re-plating. If any cracks are present, the electrical circuit cannot be completed.
13. Cloudy portion (H_3PO_4 end) is largely contaminating cell debris and can be discarded. At the anionic end (NaOH end), a thin white band will appear. This should be carefully discarded as well as it disrupts the pH gradient.
14. After allowing the fractions to settle at 4 °C in 5 mL of water, a small band of separation can be seen above the settled Sephadex. This is the liquid you want to use in your SDS-PAGE.
15. If a specific superantigen has not been identified, proceed to superantigenic fraction identification as determined by lymphocyte mitogenicity assay upon exposure to each fraction from the IEF plate.
16. Do not allow dialysis to proceed longer than 4 days. Protein will irreversibly fall out of solution.
17. Layer carefully by placing pipette tip on side and slowly allow the blood to run down the side of the tube, ideally having ~10 mL clear layer on the bottom.
18. The lymphocyte mitogenicity assay can be used on known superantigens in order to check functionality of the purified protein. Alternatively, you can perform a mitogenicity assay on unknown, potential superantigens, in which you would test the individual fractions from the IEF purification process and look for tritiated thymidine uptake into DNA by PBMCs as an indication of superantigen-stimulated lymphocyte division.
19. Fractions with lymphocyte mitogenic activity will appear as higher readings on the scintillation counter. As many dilutions are used per fraction, it should be relatively clear what fraction the superantigen is located in, as more activity at more dilute levels of supernate indicate higher concentrations of superantigen.
20. As you plunge the fluid in and out of the syringe, the protein and emulsifying agent will begin to turn opaque and become difficult to draw into the syringe. It is important to keep the mixture moving, regardless of difficulty until it is injected into the rabbit.

References

1. Argudin MA, Mendoza MC, Rodicio MR (2010) Food poisoning and *Staphylococcus aureus* enterotoxins. Toxins (Basel) 2:1751–1773
2. Altemeier WA, Lewis S, Schlievert PM, Bjornson HS (1981) Studies of the staphylococcal causation of toxic shock syndrome. Surg Gynecol Obstet 153:481–485
3. Salgado-Pabon W, Breshears L, Spaulding AR, Merriman JA, Stach CS, Horswill AR, Peterson ML, Schlievert PM (2013) Superantigens are critical for *Staphylococcus aureus* infective endocarditis, sepsis, and acute kidney injury. mBio 4:e00494–13
4. Wilson GJ, Seo KS, Cartwright RA, Connelley T, Chuang-Smith ON, Merriman JA, Guinane CM, Park JY, Bohach GA, Schlievert PM, Morrison WI, Fitzgerald JR (2011) A novel core genome-encoded superantigen contributes to lethality of community-associated MRSA necrotizing pneumonia. PLoS Pathog 7:e1002271
5. Kusch H, Engelmann S (2014) Secrets of the secretome in *Staphylococcus aureus*. Int J Med Microbiol 304:133–141
6. Schlievert PM (1982) Enhancement of host susceptibility to lethal endotoxin shock by staphylococcal pyrogenic exotoxin type C. Infect Immun 36:123–128
7. McCormick JK, Schlievert PM (2003) Expression, purification, and detection of novel streptococcal superantigens. Methods Mol Biol 214:33–43
8. Spaulding AR, Lin YC, Merriman JA, Brosnahan AJ, Peterson ML, Schlievert PM (2012) Immunity to *Staphylococcus aureus* secreted proteins protects rabbits from serious illnesses. Vaccine 30:5099–5109
9. Spaulding AR, Salgado-Pabon W, Merriman JA, Stach CS, Ji Y, Gillman AN, Peterson ML, Schlievert PM (2014) Vaccination against *Staphylococcus aureus* pneumonia. J Infect Dis 209:1955–1962
10. Schlievert PM, Shands KN, Dan BB, Schmid GP, Nishimura RD (1981) Identification and characterization of an exotoxin from *Staphylococcus aureus* associated with toxic-shock syndrome. J Infect Dis 143:509–516

Chapter 3

Strain Discrimination of *Staphylococcus aureus* Using Superantigen Profiles

Hau-Yang Tsen, Sheng-Chih Li, Yu-Cheng Chiang, and Shuo-Wen Tsai

Abstract

Staphylococcus aureus is one of the major bacterial species that may cause clinical infection and food-poisoning cases. Strains of this species may produce a series of superantigens (SAgs). Due to the importance of staphylococcal infections, reliable methods for the discrimination of strains of this species are important. Such data may allow us to trace the infection origins and be used for epidemiological study. For strain discrimination, genotyping methods, such as pulsed-field gel electrophoresis (PFGE), random amplified polymorphic DNA (RAPD), and multi-locus sequence typing (MLST), etc., could be used. Recently, toxin gene profiles, which can be used for the elucidation of the genetic and pathogenic relatedness between strains, also have been used to improve the strain discrimination. For *S. aureus*, as more SAg genes were discovered, the SAg profiles become more useful for the strain discrimination of *S. aureus*. In this chapter, a method for the discrimination of *S. aureus* strains using superantigen profiles will be described in detail.

Key words *Staphylococcus aureus*, Strain discrimination, Staphylococcal enterotoxins, Genotyping, Superantigen profiles

1 Introduction

Staphylococcus aureus is one of the major bacterial pathogens that may cause food-poisoning cases. It is also an opportunistic pathogen to cause staphylococcal infection [1, 2]. Production of different extracellular proteins with enterotoxigenicity or superantigenic activity (i.e. staphylococcal enterotoxins [SEs], Staphylococcal enterotoxin-like [SEl] toxins or termed as staphylococcal superantigens (SAgs) and toxic shock syndrome toxin [TSST-1]) has been reported for strains of this bacterial species [3, 4]. For SAgs, in addition to the classical enterotoxin types, such as SEA through SEE and TSST-1, many new types of superantigens (SAgs) (i.e. SEG, SEH, SEI, SE*l*J, SE*l*K, SElL, SE*l*M, SE*l*N, SE*l*O, SE*l*P, SE*l*Q, SER, SE*l*U, SE*l*S, SE*l*T, SE*l*V, SE*l*W, and SE*l*X) have been reported [4–6] (Table 1).

Amanda J. Brosnahan (ed.), *Superantigens: Methods and Protocols*, Methods in Molecular Biology, vol. 1396,
DOI 10.1007/978-1-4939-3344-0_3, © Springer Science+Business Media New York 2016

Due to the importance of staphylococcal infections, reliable methods for the discrimination of strains of this species are required. Such data may allow us to trace the infection origins and be used for epidemiological study. For strain discrimination, methods, such as pulsed-field gel electrophoresis (PFGE), random amplified polymorphic DNA (RAPD), and multi-locus sequence typing (MLST), etc., are frequently used. Recently, toxin gene profiles, which can be used for the elucidation of the genetic and pathogenic relatedness between strains, also have been used to improve the strain discrimination. For *S. aureus*, as more SAg genes are discovered, the SAg profile may become more useful for the strain discrimination.

Since some SAg genes are located in genetic elements, such as plasmids and prophages (Table 1), it should be mentioned that although horizontal transfer of mobile genetic elements containing virulence genes between *S. aureus* strains may lead to the emergence of many different toxin gene combinations in *S. aureus* cells during evolution [8, 9], reports have shown that transfer of these virulence

Table 1
General properties of SEs and SEls and genomic location of the encoding genes

Genomic location of the encoding genes of SAg[a]														
Enterotoxingene cluster (egc) Chromosome			Plasmid			Prophage			Pathogenicity island			Transposon		
SAg	Mwt[b]	Ema[c]	SAg	Mwt	Ema	SAg	Mwt	Ema	SAg	Mwt	Ema	SAg	Mwt	Ema
SEB	28,336	+	SEC1, 2, 3	≈		SEA	27,100	+	SEB	28,336	+	SEH	25,210	+
SEG	27,043	+		27,500	+	SEB	28,336	+	SElK	25,539	nk			
SEI	24,928	+	SED	26,360	+	SEE	26,425	+	SElL	24,593	–			
SElM	24,842	nk	SElJ	28,656	nk[d]	SElP	26,608	nk	SElQ	25,076	–			
SElN	26,067	nk	SER	27,049	+									
SElO	26,777	nk	SES	26,217	+									
SElU	27,192	nk	SET	22,614	+									
SElU2	26,672	nk												
(SEW)														
SElV	24,997	nk												
SElX	19,343	nk												

For detail, please refer to Hennekinne et al. [7]

[a]All the strains are positive for superantigenic action

[b]Mwt for superantigen (kDa)

[c]Emetic activity (Ema) demonstrated in rabbits (SElL) or in the small insectivore *Suncus murinus* but not in a primate model

[d]*nk* not known

genes is infrequent and the virulence gene expression pattern in *S. aureus* is stable [8, 10]. Several studies have characterized SAg genes in methicillin-resistant *S. aureus* (MRSA) for epidemiological investigation [8, 11, 12]. Distribution of staphylococcal toxin genes in food-poisoning isolates also has been reported. Using the SAg gene profiles, combined with PFGE, RAPD or MLST, the strain discriminatory power for *S. aureus* could be enhanced [9, 13]. Table 2 shows an example of using the SAg profiles to improve the strain discrimination

Table 2
Subtyping by pulsed-field gel electrophoresis (PFGE) of the *SmaI*-digested chromosomal DNA and superantigen profiles for 73 clinical isolates of *Staphylococcus aureus*

PFGE	Superantigen profiles
C1	sea (1), sea-sek (1), sea-sep (1), sea-tst (1), sea-sek-seq-tst (2), sea-sei-seq (1), sea-sel-sem-seq-tst (1), sea-sep-seq (1), sea-seq (4), sep (1), seq (2)
C2	sea (1), sea-sek (1), sea-tst (1), sea-seq (4), sea-sei-sem-sep-seq (1), sea-sek-seq-tst (1), sea-sep-seq (2)
C3	sep-tst (1)
C4	sea-seq (1), sec-seg-sek-sel-sem-sen-seo-sep-seu (1), sec-seg-sei-sek-sel-semsen-seo-sep-seu-tst (1), tst (1)
C5	sea-sep-seq (1), seb-seq (1), seb-sek-seq-tst (1)
C6	ND (1)
C7	seb-sei-seq-tst (1), sep-tst (1),sep (1)
C8	sep (1)
C9	sei-sep (1), ND (1)
C10	seb-sek-seq (1), seb-seq (1), seb-seo-seq (1)
C11	sea-seh-sei-ser-tst (1), seb-sei-tst (1), sep (1)
C12	ND (1)
C13	sep (2), ND (2)
C14	sea-sek-sep-seq (1), sea-sep (1), seh-sep-tst (1)
C15	sec-sei-sek-sel-semsen-seo-sep-seu-tst (2), sen (1), sen-seo-seu (1)
C16	seb-sep-seq (1), sep (2)
C17	sec-sei-sel-sem-senseo-tst (1), tst (1), ND (2)
C18	sec-seg-sei-sek-selsem-sen-seo-sep-tst (1), sec-seg-sei-sek-sel-semsen-seo-sep-seu-tst (1), sec-seg-sei-sek-sel-semsen-seo-sep-seu (1)
C19	seb-seo-seu (1),sec-seg-sei-sek-sel-semsen-seo-seu-tst (1)
C20	tst (1)

Total: 20 PFGE clusters (with cutoff at 52 % similarity), 59combined patterns
The numbers in parentheses indicate the number of strains with the same PFGE pattern and the same SAg profile
ND: Superantigen genes assayed were not detected
For detail, please refer to Chiang et al. [13]

of *S. aureus*. Strains in the same PFGE pattern could be further discriminated by SAg typing. Only when several genotyping methods, such as PFGE and RAPD, were combined for strain typing, and using high strain similarity (>80 %) cutline for dendrogram analysis, strains may not be further discriminated by SAg profiles. Nowadays, due to the discovery of more superantigen genes for *S. aureus*, the SAg profiles become more powerful for the discrimination of *S. aureus* strains.

To establish the SAg profiles for *S. aureus* strains, it is important to keep in mind that each individual strain may harbor more than one SAg genes. Therefore, for each strain, all the known SAg genes should be assayed. For such purpose, it is possible to use the multiplex PCR system which allows the simultaneous detection of two or more, but not too many of the SAg genes. A multiplex system specific for different SAg genes has been reported [14]. In this chapter, the experimental conditions and the primers for the determination of all the known SE and SAg gene in *S. aureus* strains are described.

2 Materials

2.1 Bacterial Strains

1. Reference enterotoxigenic strains available in Bioresource Collection and Research Center (BCRC), Hsin-Chu, Taiwan, or other culture collection centers can be used.
2. Strains previously characterized as superantigenic *S. aureus* strains can be used as reference strains. For example, strains reported by Chiang et al. [15, 16].
3. Species of *S. aureus* isolates from different origins, such as from food-poisoning cases, clinical samples or other laboratories should be identified according to the methods described in Bacteriological Analytical Manual [17], or checked by PCR using *S. aureus*-specific primers, such as those reported by Chiang et al. [18].

2.2 Media and Reagents for DNA Preparation

1. Trypticase soy broth (TSB): Enzymatic Digest of Casein 17.0 g, Enzymatic Digest of Soybean Meal 3.0 g, Sodium Chloride 5.0 g, Dipotassium Phosphate 2.5 g, Dextrose 2.5 g or use the commercial TSB product 30 g (Becton, Dickinson and Company, Sparks, MD, USA).
   (a) Dissolve the 30 g TSB medium (or individual ingredients) in 1 L ddH_2O and mix thoroughly.
   (b) Autoclave at 121 °C for 15 min.
2. Brain heart infusion medium: Brain Heart Infusion 17.5 g, Enzymatic Digest of Gelatin 10 g, Dextrose 2 g, Sodium

Chloride 5 g, Disodium Phosphate 2.5 g, or use the commercial BHI product 37 g (Becton, Dickinson and Company).

(a) Dissolve 37 g of the medium (or individual components) in 1 L of ddH_2O.

(b) Heat with frequent agitation to completely dissolve the medium.

(c) Autoclave at 121 °C for 15 min.

3. PIV buffer: 1 mol/L NaCl, 10 mmol/L Tris–HCl, pH 7.6.
4. Lysostaphin buffer: 6 mM Tris–HCl, pH 7.6, 1 M NaCl, 100 mM EDTA, 0.5 % Brij 58, 0.2 % deoxycholate.
5. Lysostaphin, Bio Ultra, Activity ≥2000 U/mg (Sigma-Aldrich®, USA).
6. Lysozyme, Activity > 20,000 U/mg, 10 g.
7. RNase A, Activity > 60 U/mg, 100 mg.
8. Proteinase K, Activity > 60 U/mg.
9. Tris buffer saturated phenol (pH 8.0).
10. Chloroform solution.
11. 99.5 % ethanol.
12. Phenol–chloroform (1:3) solution.
13. TE buffer: 10 mmol/L Tris–HCl, pH 8.0, 1 mmol/L EDTA.
14. Blood and Tissue Genomic DNA Miniprep System (Viogene, Hsinchu, Taiwan).

2.3 Reagents and Stock Solutions for PCR

1. RNase-free water or distilled deionized water (ddH_2O).
2. 10× Tris-triton buffer: 100 mM Tris–HCl, pH 8.8 at 25 °C, 15 mM $MgCl_2$, 500 mM KCl, 1 % Triton X-100.
3. dNTPs (10 mM) or dATP (100 mM), dTTP (100 mM), dCTP (100 mM), dGTP (100 mM).
4. Primers mixed with ddH_2O to 100 mM according to the instruction of manufacturer. The sequences of PCR primers targeted to each SAg gene and the PCR conditions including the reaction temperature and time for each PCR step (i.e. template denaturation, primer annealing, and chain extension) for each SAg gene are shown in Table 3 (*see* **Note 1**).
5. Taq DNA polymerase 2 U/μl.

2.4 Reagents for Agarose Gel Electrophoresis

1. Tris-acetate-EDTA (TAE) buffer, 50×.

(a) Take 242 g Tris Base (MW = 121.1), and mix with ddH_2O to 600 ml followed by stirring.

(b) Add 57.1 mL Glacial Acetic Acid and 100 ml 0.5 M EDTA.

Table 3
PCR primers and conditions for the detection of *S. aureus* and its enterotoxin genes

Target gene	Primer set	Sequence of primers	Condition	Predicted size of product (bp)	References
S. aureus htrA gene	SalS[a] SalA	5′-ACAAATAATAAAGGTGGC-3′/(F) 5′-GCATGTTAATAACTCCAA-3′(R)	94 °C/20 s 52 °C/20 s 72 °C/60 s	109	[18]
sea	**SEA1**[b] **SEA2**	5′-AAAGTGCCGATCAATTTATGGCTA-3′/(F) 5′-GTAATTAACCGAAGGTTCTGTAGA-3′(R)	94 °C/20 s 51.5 °C/20 s 72 °C/20 s	219	[19]
seb	**SEB1** **SEB2**	5′-TCGCATCAAACTGACAAACGA-3′/(F) 5′-CACTTTTTCTTTGTCGTAAGATAA-3′ (R)	94 °C/20 s 54 °C/20 s 72 °C/20 s	410	[20]
sec	**SEC1** **SEC2**	5′-AACATTAGTGATAAAAAAGTGAAA-3′/(F) 5′-TTGTAAGTTCCCATTATCAAAGTG-3′ (R)	94 °C/20 s 54 °C/20 s 72 °C/20 s	234	[20]
sed	**SED1** **SED2**	5′-GCAGATAAAAATCCAATAATAGGA-3′/(F) 5′-TACTAAAGAAACTTCTTTTTGTAC-3′(R)	94 °C/20 s 54 °C/20 s 72 °C/20 s	331	[19]
see	**SE1** **SE2**	5′-TTACAAAGAAATGCTTTAAGC-3′/(F) 5′-TAAACCAAATTTTCCGTG-3′(R)	94 °C/30 s 48 °C/30 s 72 °C/30 s	456	[19]
seg	**SEG1** **SEG2**	5′-GCTATCGACACACTACAACC-3′/(F) 5′-CCAAGTGATTGTCTATTGTCG-3′(R)	94 °C/20 s 48 °C/20 s 72 °C/20 s	583	[21]
seh	**SEH1** **SEH2**	5′-CACATCATATGCGAAAGC-3′/(F) 5′-CGAATGAGTAATCTCTAGG-3′(R)	94 °C/30 s 58 °C/30 s 72 °C/30 s	548	[21]

sei	**SEI1** **SEI2**	5′-GATACTGGAACAGGACAAGC-3′/(F) 5′-CTTACAGGCAGTCCATCTCC-3′ (R)	94 °C/30 s 58 °C/30 s 72 °C/30 s	789	[21]
sej	**SEJ1** **SEJ2**	5′-CTCCCTGACGTTAACACTACTAATAA-3′/(F) 5′-TTGTCTGGATATTGACCTATAACATT-3′/(R)	94 °C/30 s 55 °C/30 s 72 °C/30 s	666	[22]
sek	SEK1 SEK2	5′-CACAGCTACTAACGAATATC-3′/(F) 5′-TGGAATTTCTCAGACTCTAC-3′(R)	94 °C/20 s 56 °C/20 s 72 °C/25 s	378	[15]
sel	SEL1 SEL2	5′-CATACAGTCTTATCTAACGG-3′/(F) 5′-TTTTCTGCTTTAGTAACACC-3′ (R)	94 °C/20 s 56 °C/20 s 72 °C/20 s	275	[15]
sem	SEM1 SEM2	5′-CTTGTCCTGTTCCAGTATC-3′/(F) 5′-ATACGGTGGAGTTACATTAG-3′ (R)	94 °C/20 s 56 °C/20 s 72 °C/20 s	329	[15]
sen	SEN1 SEN2	5′-CTTCTTGTTGGACACCATCTT-3′/(F) 5′-GAAATAAATGTGTAGGCTT-3′/(R)	94 °C/30 s 55 °C/30 s 72 °C/20 s	135	[18]
seo	SEO1 SEO2	5′-AAATTCAGCAGATATTCCAT-3′/(F) 5′-TTTGTGTAAGAAGTCAAGTGTAG-3′/(R)	94 °C/30 s 56 °C/30 s 72 °C/20 s	172	[18]
sep	SEP1 SEP2	5′-ATCATAACCAACCGAATCAC-3′/(F) 5′-AGAAGTAACTGTTCAGGAGCTA-3′/(R)	94 °C/30 s 55 °C/30 s 72 °C/20 s	148	[18]
seq	SEQ1 SEQ2	5′-TCAGGTCTTTGTAATACAAAA-3′/(F) 5′-TCTGCTTGACCAGTTCCGGT-3′/(R)	94 °C/30 s 55 °C/30 s 72 °C/30 s	359	[18]
ser	SER1 SER2	5′-AGATGTGTTTGGAATACCCTAT-3′/(F) 5′-CTATCAGCTGTGGAGTGCAT-3′/(R)	94 °C/30 s 55 °C/30 s 72 °C/30 s	123	[18]

(continued)

Table 3 (continued)

Target gene	Primer set	Sequence of primers	Condition	Predicted size of product (bp)	References
seu	SEU1 SEU2	5′-ATTTGCTTTTATCTTCAT-3′/(F) 5′-GGACTTTAATGTTTGTTTCTGAT-3′/(R)	94 °C/30 s 55 °C/30 s 72 °C/30 s	167	[18]
ses	ses-ORF6F ses-ORF6R	5′-TTCAGAAATAGCCAATCATTTCAA-3′/(F) 5′-CCTTTTTGTTGAGAGCCGTC-3′/(R)	94 °C/30 s 51 °C/30 s 72 °C/30 s	195	[23]
set	set-ORF5F set-ORF5R	5′-GGTGATTATGTAGATGCTTGGG-3′/(F) 5′-TCGGGTGTTACTTCTGTTTGC-3′/(R)	94 °C/30 s 55 °C/30 s 72 °C/30 s	170	[23]
se*l*w(se*l*u2)[d]	R-SElU2-1 R-SElU2-2	5′-CAGGATCCATGTTAAATGGCAATCCTAAACCA-3′/(F) 5′-GCCTGCAGTTATTTTTTGGTTAAATGAACTTCTACATT AATAGATTTA-3/(R)	94 °C/30 s 56 °C/30 s 72 °C/30 s	720	[24]
se*l*v	R-SElV 1[c] R-SElV 2	5′-GCAGGATCCGATGTCGGAGTTTTGAATCTTAGG-3′/(F) 5′-TAACTGCAGTTAGTTACTATCTACATATGATATTTCGACATC-3/ (R)	94 °C/30 s 51 °C/30 s 72 °C/30 s	653	[24]
se*l*x	*Selx*seq F *Selx*seq R	5′-TGGTAGCAAATTAAAGTTAATCAAGAG-3′/(F) 5′-TGCTAATCATAACAAAGAAAGCTAGG-3/(R)	94 °C/30 s 56 °C/30 s 72 °C/30 s	825	[25]
tsst	**TSST1**[e] **TSST2**	5′-ATGGCAGCATCAGCTTGATA-3′/(F) 5′-TTTCCAATAACCACCCGTTT-3′ (R)	94 °C/30 s 55 °C/30 s 72 °C/30 s	350	[26]

[a]staphylococcus aureus specific primers
[b]*SE* Staphylococcal enterotoxin
[c]*SEl* Staphylococcal enterotoxin-like
[d]se*l*w also termed as se*l*u2
[e]*TSST* Toxic shock syndrome toxin

(c) Bring final volume to 1 L with ddH_2O.
(d) Store at room temperature.
(e) Dilute 1:50 for 1× concentration for use.

2. Ethidium bromide 10 mg/ml.
3. Agarose I.
4. 6× Loading buffer which consists of 30 % (W/V) Glycerol, 0.25 % (W/V) Bromophenol blue, and 0.25 % (W/V) Xylene cyanol.

2.5 Instruments

1. Dry-bath.
2. PCR Thermal Cycler.
3. Electrophoresis system.
4. UV box.
5. Gel Documentation system.
6. Centrifuge.
7. Shaker.
8. Vortex mixer.

3 Methods

3.1 Cell Cultivation and Preparation of Staphylococcal DNA

1. Culture bacterial cells in TSB or brain heart infusion and incubate at 37 °C for 8–12 h (*see* **Note 2**).
2. Collect 3 ml of bacterial cells from 12 h culture in TSB or BHI and wash with 1 ml PIV buffer.
3. Pellet cells and suspend in 0.4 ml lysostaphin buffer.
4. Add 20 μl of lysostaphin (2 mg/ml), 20 μl lysozyme (2 mg/ml), and 20 μl RNase (2 mg/ml), and incubate at 37 °C for 1 h.
5. Add 25 μl proteinase K (10 mg/ml) and incubate at 60 °C for 1 h.
6. Add an equal volume of Tris buffer (pH 8.0) saturated phenol.
7. Invert the mixture gently 3–5 times.
8. Centrifuge the mixture at 12,000 ×*g* for 10 min.
9. Mix supernatant with an equal volume of phenol–chloroform, invert gently for 3–5 times, and centrifuge the mixture at 12,000 ×*g* for 10 min.
10. Mix the supernatant again with an equal volume of chloroform, invert gently for 3–5 times and centrifuge the mixture at 12,000 ×*g* for 10 min.

11. DNA in the supernatant is precipitated by mixing with 2 volumes of 95 % ethanol, followed by storage at −70 °C for 1 h.
12. Wash the DNA pellet with ice-cold 70 % ethanol.
13. Centrifuge to remove ethanol.
14. Vacuum dry the DNA pellet.
15. Suspend DNA in sterile TE buffer and store at 4 °C (*see* **Note 3**).

3.2 Preparation of the PCR Premix and Reaction Mixture

S. aureus strain may harbor several different SE or SAg genes, thus for each strain all the SE or SAg genes should be assayed. Multiplex PCR, which allows the simultaneous detection of more than one SE or SAg genes, may be used. However never try to detect more than 5 SAg genes simultaneously, otherwise some SAg genes may not be detectable. When performing multiplex PCR, be sure to select the primer sets that require similar PCR conditions, but still generate products with distinguishable molecular weights.

3.2.1 Preparation of Premix

1. Dilute the stock solutions listed in Subheading 2.3 with ddH_2O. For dNTPs (2.5 mM), 1 ml of 10 mM stock dNTPs was mixed with ddH_2O to 4 ml; or 250 μl of each of the 100 mM stocks of dATP, dTTP, dCTP and dGTP was mixed and diluted with 9 ml ddH_2O to 10 ml.
2. For primers, 10 μl of the stock was mixed with 90 μl ddH_2O.
3. In a 200 μl eppendorf tube, pipet 16.3 μl of RNase-free water or ddH_2O, then mix with 2.5 μl 10× Tris-triton buffer; 2.0 μl of dNTPs, 1 μl each of the forward and reverse primers, and 0.2 μl Taq DNA polymerase (2U/μl). The final mixture is called PCR premix (*see* **Note 4**).

3.2.2 PCR Reaction Mixture and Reaction

1. Add 2 μl DNA sample to the PCR premix and place the tube in PCR Thermal Cycler (*see* **Note 5**).
2. Use the following PCR cycle: 94 °C for 5 min followed by 35 cycles of template denaturation, primer annealing, and chain extension. The specific cycling temperatures and times are given under *Condition* in Table 3 for each SAg gene (based on their specific primer sets). After amplification, a final extension step of 7 min at 72 °C was performed for all PCR reactions.
3. The molecular weights of amplified PCR products were than analyzed by gel electrophoresis as described in Subheading 3.3 (*see* **Note 6**).

3.3 Agarose Gel Electrophoresis

3.3.1 Preparation of 2 % Agarose Gel

1. Measure 6 g agarose powder and add it to a 500 ml flask.
2. Add 300 ml 1× TAE Buffer to the flask (the total gel volume can be varied depending on the size of the casting tray).

3. Melt the agarose in a microwave or hot water bath until the solution becomes clear (*see* **Note 7**).
4. Let the solution cool to about 50–55 °C, swirl the flask occasionally to cool evenly.
5. Seal the ends of the casting tray with two layers of tape.
6. Place the combs in the gel casting tray.
7. Pour the melted agarose solution into the casting tray and let cool until it is solid (it should appear milky white).
8. Carefully pull out the combs and remove the tape.
9. Place the gel in the electrophoresis chamber.
10. Add enough 1× TAE Buffer so that there is about 2–3 mm of buffer over the gel.

3.3.2 Loading and Running the Gel

1. After PCR, mix 1 μl of 6× Sample loading buffer with 3–5 μl PCR product or DNA standard or marker.
2. Carefully pipette 4 μl of the above loading mixture into each well in the gel.
3. Place the lid on the gel box, connecting the electrodes.
4. Turn on the power supply to approximately 50 V and run for 3–5 min.
5. Check to make sure the current is running through the buffer by looking for bubbles forming on each electrode.
6. Turn up the power supply to 100 V for 30–35 min. Let the power run until the blue dye approaches the end of the gel.
7. Turn off the power.
8. Remove the lid of the electrophoresis chamber.
9. Using gloves, carefully remove the tray and gel.

3.3.3 Gel Staining

1. Prepare enough volume of 0.5 μg/ml ethidium bromide solution in ddH_2O to completely submerge the gel (*see* **Note 8**).
2. After electrophoresis, submerge the gel in the staining solution for 15–30 min (depending upon gel thickness).
3. Place the gel on plastic wrap on a UV light box and observe under 300 nm illumination. Bands will appear bright orange on a pale orange background.
4. Use Gel Documentation System to take photo and analyze the results.
5. An example for the PCR detection of SEQ and SEU genes from *S. aureus* strains are shown in Figs. 1 and 2. Clearly, cell counts were very important for the positive detection of the SEQ and SEU genes. With cell counts less than 10^1 CFU/ml, a 10 h preincubation would be required to generate positive result.

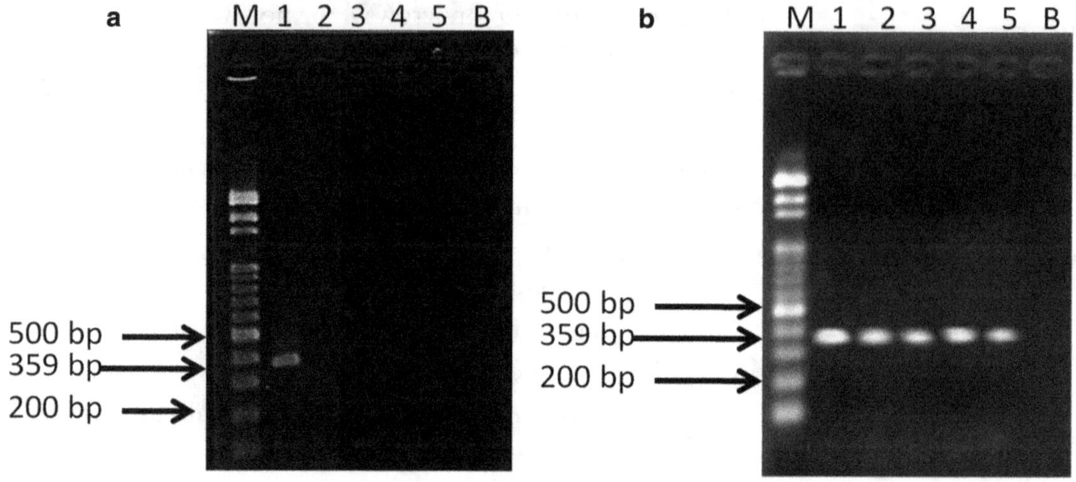

Fig. 1 PCR sensitivity for the amplification of the SEQ gene in laboratory isolate of *S. aureus* with primers SEQ1/SEQ2. The molecular size amplified was 359 bp. (**a**) Cell counts used in *lanes 1–5* were $N \times 10^4$, 10^3, 10^2, 10^1, 10^0 CFU/ml, respectively. (**b**) Cell counts used in *lanes 1–5* were the 10 h culture in TSB of those shown in (**a**), respectively. *B* Blank

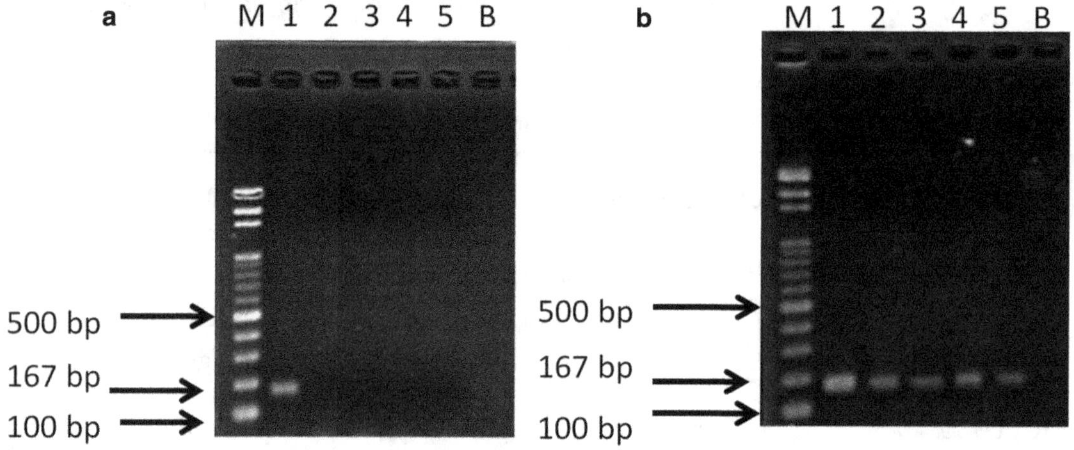

Fig. 2 PCR sensitivity for the amplification of the SEU gene in laboratory isolate of *S. aureus* with primers SEU1/SEU2. The molecular size amplified was 167 bp. (**a**) Cell counts used in *lanes 1–5* were $N \times 10^4$, 10^3, 10^2, 10^1, 10^0 CFU/ml, respectively. (**b**) Cell counts used in *lanes 1–5* were the 10 h culture in TSB of those shown in (**a**), respectively. *B* Blank

4 Notes

1. According to the gene sequences available for SAg in GeneBank, the primers for each of the SAg genes may be self-designed. However, one must first check the specificity of these primers by PCR assay using reference strains with known SAg genes, or by sequencing the amplified product and comparing the sequence with this sequence available in GeneBank.

2. The detection limit for PCR assay of the SE or SAg genes in *S. aureus* cells is in general >10^3 CFU/ml. Therefore, to obtain clear results, DNA from 10^3 CFU/ml of *S. aureus* cells is needed.
3. For PCR detection of SE or SAg genes, the concentration of template DNA should be within the range between 50 and 200 ng/μl (The final concentrations may range from 4 to 16 ng/μl); lower than 30 ng/μl may lead to false negative result.
4. If large quantity of samples are to be assayed, be sure to keep the PCR Master mix and reagents under cooling conditions (0–4 °C) to ensure that the enzyme and reagents are not inactivated.
5. The final concentration of each component in the PCR mixture will be 1× buffer; 0.016 U/μl Taq DNA polymerase; 200 μM dNTPs; 0.4 μM of each primer, and 4–16 ng/μl of DNA sample. For primers, the concentration could be within the range of 0.05–0.4 μM.
6. The PCR products obtained for SAg genes may be confirmed by sequencing the products and aligning sequences with the published SAg sequence.
7. If using a microwave, heat the agarose solution for several short intervals—do not let the solution boil for long periods as it may boil out of the flask.
8. The ethidium bromide solution is stable for 1–2 months at room temperature in the dark.

Acknowledgement

We thank very much to the Ministry of Science and Technology, Taipei, Taiwan for the support of all the works required for this manuscript in the past years. The recent project for this work is NSC 102-2632-B-241-001-MY3.

References

1. Lina G, Pie'mont Y, Godail-Gamot F, Bes M, Peter MO, Gauduchon V, Vandenesch F, Etienne J (1999) Involvement of Panton-Valentine leukocidin-producing *Staphylococcus aureus* in primary skin infections and pneumonia. Clin Infect Dis 29:1128–1132
2. Gill SR, Fouts DE, Archer GL, Mongodin EF, Deboy RT, Ravel J, Paulsen IT, Kolonay JF, Brinkac L et al (2005) Insights on evolution of virulence and resistance from the complete genome analysis of an early methicillin-resistant *Staphylococcus aureus* strain and a biofilm-producing methicillin resistant *Staphylococcus epidermidis* strain. J Bacteriol 187:2426–2438
3. Balaban N, Rasooly A (2000) Staphylococcal enterotoxins. Int J Food Microbiol 61:1–10
4. Hu DL, Nakane A (2014) Mechanisms of staphylococcal enterotoxin induced emesis. Eur J Pharmacol 722:95–107

5. Le Loir Y, Baron F, Gautier M (2003) *Staphylococcus aureus* and food poisoning. Genet Mol Res 2:63–76
6. Omoe K, Hu DL, Takahashi-Omoe H, Nakane A, Shinagawa K (2005) Comprehensive analysis of classical and newly described staphylococcal superantigenic toxin genes in *Staphylococcus aureus* isolates. FEMS Microbiol Lett 246:191–198
7. Hennekinne JA, Ostyn A, Guillier F, Herbin S, Prufer AL, Dragacci S (2010) How should staphylococcal food poisoning outbreaks be characterized? Toxins (Basel) 2:2106–2116. doi:10.3390/toxins2082106
8. Diep BA, Carleton HA, Chang RF, Sensabaugh GF, Perdreau-Remington F (2006) Roles of 34 virulence genes in the evolution of hospital- and community-associated strains of methicillin resistant *Staphylococcus aureus*. J Infect Dis 193:1495–1503
9. Xie Y, He Y, Gehring A, Hu Y, Li Q, Tu SI, Shi X (2011) Genotypes and toxin gene profiles of *Staphylococcus aureus* clinical isolates from China. PLoS One 6:e28276
10. Ziebandt AK, Kusch H, Degner M, Jaglitz S, Sibbald MJ, Arends JP, Chlebowicz MA, Albrecht D, Pantucek R et al (2010) Proteomics uncovers extreme heterogeneity in the *Staphylococcus aureus* exoproteome due to genomic plasticity and variant gene regulation. Proteomics 10:1634–1644
11. Coia JE, Browning L, Haines L, Birkbeck TH, Platt DJ (1992) Comparison of enterotoxins and haemolysins produced by methicillin-resistant (MRSA) and sensitive (MSSA) *Staphylococcus aureus*. J Med Microbiol 36:164–171
12. Schmitz FJ, MacKenzie CR, Geisel R, Wagner S, Idel H, Verhoef J, Hadding U, Heinz HP (1997) Enterotoxin and toxic shock syndrome toxin-1 production of methicillin resistant and methicillin sensitive *Staphylococcus aureus* strains. Eur J Epidemiol 13:699–708
13. Chiang YC, Lai CH, Lin CW, Chang CY, Tsen HY (2014) Improvement of strain discrimination by combination of superantigen profiles, PFGE, and RAPD for *Staphylococcus aureus* isolates from clinical samples and food-poisoning cases. Foodborne Pathog Dis 11:468–477. doi:10.1089/fpd.2013.1708
14. Bl A, Zdenkova K, Sykorova H, Demnerova K (2014) Molecular analysis of *Staphylococcus aureus* pathogenicity islands (SaPI) and their superantigens combination of food samples. J Microbiol Methods 107:197–204
15. Chiang YC, Chang LT, Lin CW, Yang CY, Tsen HY (2006) PCR primers for the detection of Staphylococcal enterotoxins (SEs) K, L, M and survey of SEs types in *Staphylococcus aureus* isolates from food-poisoning cases in Taiwan. J Food Prot 69:1072–1079
16. Chiang YC, Liao WW, Fan CM, Pai WY, Chiou CS, Tsen HY (2008) PCR detection of Staphylococcal enterotoxins (SEs) N, O, P, Q, R, U, and survey of SE types in *Staphylococcus aureus* isolates from food-poisoning cases in Taiwan. Int J Food Microb 121:66–73
17. Food and Drug Administration (1995) Bacteriological analytical manual, 8th edn. AOAC International, Rockville, MD
18. Chiang YC, Fan CM, Liao WW, Lin CK, Tsen HY (2007) Real-time PCR detection of *Staphylococcus aureus* in milk and meat using new primers designed from the heat shock protein gene htrA sequence. J Food Prot 70:2855–2859
19. Tsen HY, Chen TR (1992) Use of PCR for the specific detection of type A, D, E enterotoxigenic *S. aureus* in food. Appl Microbiol Biotechnol 37:685–690
20. Tsen HY, Chen TR, Yu GY (1994) Detection of B and C types enterotoxigenic *Staphylococcus aureus* using polymerase chain reaction. J Chinese Agric Chem Sci 32:322–331
21. Chen TR, Chiou CC, Tsen HY (2004) Use of novel PCR primers specific to the genes of staphylococcal enterotoxin G, H, I for the survey of *Staphylococcus aureus* strains isolated from food-poisoning cases and food samples in Taiwan. Int J Food Microbiol 92:189–197
22. Becker K, Friedrich AW, Lubritz G, Weilert M, Peters G, von Eiff C (2003) Prevalence of genes encoding pyrogenic toxin superantigens and exfoliative toxins among strains of *Staphylococcus aureus* isolated from blood and nasal specimens. J Clin Microbiol 41:1434–1439
23. Ono HK, Omoe K, Imanishi K, Iwakabe Y, Hu DL, Kato H, Saito N, Nakane A, Uchiyama T, Shinagawa K (2008) Identification and characterization of two novel staphylococcal enterotoxins, types S and T. Infect Immun 76:4999–5005. doi:10.1128/IAI.00045-08
24. Thomas DY, Jarraud S, Lemercier B, Cozon G, Echasserieau K, Etienne J, Gougeon ML, Lina G, Vandenesch F (2006) Staphylococcal enterotoxin-like toxins U2 and V, two new staphylococcal superantigens arising from recombination within the enterotoxin gene cluster. Infect Immun 74:4724–4734

25. Wilson GJ, Seo KS, Cartwright RA, Connelley T, Chuang-Smith ON, Merriman JA, Guinane CM, Park JY, Bohach GA, Schlievert PM, Morrison WI, Fitzgerald JR (2011) A novel core genome-encoded superantigen contributes to lethality of community-associated MRSA necrotizing pneumonia. PLoS Pathog 7:e1002271

26. Johnson WM, Tyler SD, Ewan EP, Ashton FE, Pollard DR, Rozee KR (1991) Detection of genes for enterotoxins, exfoliative toxins, and toxic shock syndrome toxin 1 in *Staphylococcus aureus* by the polymerase chain reaction. J Clin Microbiol 29:426–430

Chapter 4

Crystallization and Structure Determination of Superantigens and Immune Receptor Complexes

Karin E.J. Rödström and Karin Lindkvist-Petersson

Abstract

Structure determination of superantigens and the complexes they form with immune receptors have over the years provided insight in their modes of action. This technique requires growing large and highly ordered crystals of the superantigen or receptor–superantigen complex, followed by exposure to X-ray radiation and data collection. Here, we describe methods for crystallizing superantigens and superantigen–receptor complexes using the vapor diffusion technique, how the crystals may be optimized, and lastly data collection and structure determination.

Key words Crystallization, Structure determination, Protein–protein complexes

1 Introduction

Superantigens (SAgs) are immune stimulatory toxins capable of inflicting serious disease in humans, such as food poisoning and toxic shock syndrome [1, 2]. Here, we focus on the staphylococcal enterotoxins, which share a common three-dimensional fold. Structural studies of these superantigens [3] and superantigen complexes [4, 5] have over the years revealed how they bind to the human receptors, the T-cell receptor (TCR), and the major histocompatibility complex (MHC) class II. In addition, they have contributed to the insight into their mechanism of action and allowed for development of potential antagonists [6–8] or vaccines against SAgs [9–11].

Since superantigens are stable, rigid molecules, they are excellent targets for crystallization, and have even been used in order to facilitate crystallization of MHC class II [12, 13]. To date, over a 100 structures of staphylococcal enterotoxins and streptococcal pyrogenic exotoxins have been determined, but these structures only represent approximately 50 % of the characterized toxins so far. Thus, there are still a number of superantigens that have not

Amanda J. Brosnahan (ed.), *Superantigens: Methods and Protocols*, Methods in Molecular Biology, vol. 1396,
DOI 10.1007/978-1-4939-3344-0_4, © Springer Science+Business Media New York 2016

been studied structurally. Although the first superantigen structure, staphylococcal enterotoxin B (SEB), was determined in 1992 [3], it took 18 years until the first complete structure of a staphylococcal enterotoxin in complex with both TCR and MHC class II was determined (SEH) [14], and another 4 years until SEB was determined in complex with both its receptors [15].

Here, we present the methodology behind crystallizing superantigens and superantigen–receptor complexes, preparation of the crystals for data collection and structure determination used for the TCR–SAg–MHC structures [14, 15]. Naturally, other groups have demonstrated successful ways of crystallizing superantigens and complexes to what is presented here [3–5, 16–20]. In addition, crystallography is not an exact science and heavily influenced by random events, there are therefore many ways to determine superantigen structures.

2 Materials

Prepare all buffers and stock solutions using ultrapure water (18 MΩ cm). The chemicals listed in Subheading 2.3 are only suggestions and are not necessarily required for carrying out crystallization, and other chemicals that are not listed may be needed in addition.

2.1 Sample Preparation Materials

1. Tris-buffered saline (TBS): 50 mM Tris pH 7.4, 150 mM NaCl. Dissolve 6.06 g Tris base and 8.77 g NaCl in 800 ml ultrapure water and adjust pH to 7.4 (*see* **Note 1**). Adjust the volume to 1000 ml and filter through a 0.22 μm filter.
2. Size exclusion column with appropriate separation range, for instance Superdex 75 10/300 GL or Superdex 200 10/300 GL.
3. Concentration tubes with molecular weight cut-off 10,000 Da.
4. Spin filters, 0.22 μm, for volumes less than 500 μl.

2.2 Commercial Crystal Screens

1. PEGRx 1 and PEGRx 2 (Hampton Research).
2. Additive screen (Hampton research).

2.3 Buffers, Salts, and Precipitants

1. 50 % (v/v) polyethylene glycol (PEG) 200, 300, 400, and 600 and polyethylene glycol monomethyl ether (PEG MME) 350 and 550 solutions: Add 25 ml of each liquid PEG or PEG MME to 25 ml ultrapure water. Filter through 0.22 μm filters and store at 4 °C.
2. 50 % (w/v) PEG 1000, 1500, 2000, 3000, 3350, 4000, 6000, 8000, and 10000 and PEG MME 2000 and 5000 solutions: Dissolve 25 g of each solid PEG or PEG MME in ultrapure water and adjust to a final volume of 50 ml. Filter through

0.22 μm filters (*see* **Note 2**) and store at 4 °C. PEG 600 has a melting temperature just above room temperature and may need to be melted before use.

3. 25 % (w/v) PEG 20000 solution: Dissolve 12.5 g PEG 20000 in ultrapure water and adjust to a final volume of 50 ml. Filter through a 0.22 μm filter and store at 4 °C.
4. 0.5 M buffer stocks: Prepare 50 ml stocks of buffers by dissolving the respective amount of buffer in 40 ml ultrapure water each, adjust to the correct pH with sodium hydroxide or hydrochloric acid, and make up the volume to 50 ml. Filter through a 0.22 μm filter. Use 3.40 g sodium acetate trihydrate pH 4.5, 7.35 g tri-sodium citrate dihydrate pH 5.0, 4.88 g 2-(*N*-morpholino) ethanesulfonic acid (MES) hydrate pH 5.5 (*see* **Note 3**), 3.00 g monosodium phosphate pH 6.0, 5.35 g sodium cacodylate trihydrate pH 6.5 (*see* **Note 4**), 5.23 g 2-[bis(2-hydroxyethyl)amino]-2-(hydroxymethyl)-1,3-propanediol (BIS-TRIS) pH 6.5, 1.70 g imidazole pH 7.0 (*see* **Note 4**), 7.56 g piperazine-*N*,*N*′-bis(2-ethanesulfonic acid) (PIPES) pH 7.0, 5.96 g 4-2-hydroxyethyl-1-piperazineethanesulfonic acid (HEPES) pH 7.5, 3.03 g tris(hydroxymethyl)methylamine (Tris) pH 8.0, 1.34 g ammonium chloride pH 8.5, and 1.88 g glycine pH 9.0.
5. 1 M NaCl stock: Dissolve 2.93 g NaCl in ultrapure water and adjust the volume to 50 ml. Filter through a 0.22 μm filter.

2.4 Crystallization Equipment

1. Upright stereo brightfield microscope, 10× magnification.
2. Sitting-drop plates with 24 wells.
3. Hanging-drop plates with 24 (18 mm) wells.
4. Crystal clear 1.88 in. sealing tape.
5. Immersion oil.
6. Siliconized 18 mm cover slides.
7. A pair of flat tweezers.
8. Pipette, 0.2–2 μl with suitable tips.

2.5 Crystal Mounting Tools

1. Liquid nitrogen canister, 312 mm in height.
2. Cryo-protectants: 98 % glycerol and 100 % PEG 400.
3. Liquid nitrogen canes.
4. Magnetic cryo caps, for 18 mm SPINE standard and magnetic cryo vials compatible with EMBL/ESRF sample changer baskets (*see* **Note 5**).
5. Mounted litho loops, 0.06 mm, 0.1 mm, 0.2 mm, and 0.3 mm, in 18 mm pins.
6. Super glue gel.

7. Cryo wand with magnetic plunger.
8. Vial Clamp, curved 110°/70°.
9. Glove liners.

3 Methods

Crystallization of a superantigen or a superantigen–receptor complex is carried out in multiple steps described below. A flow chart of the process is shown in Fig. 1.

3.1 Sample Preparation

For crystallization of superantigens, follow **steps 1, 2**, and 7. For crystallization of superantigen–receptor complexes, follow **steps 3–7**.

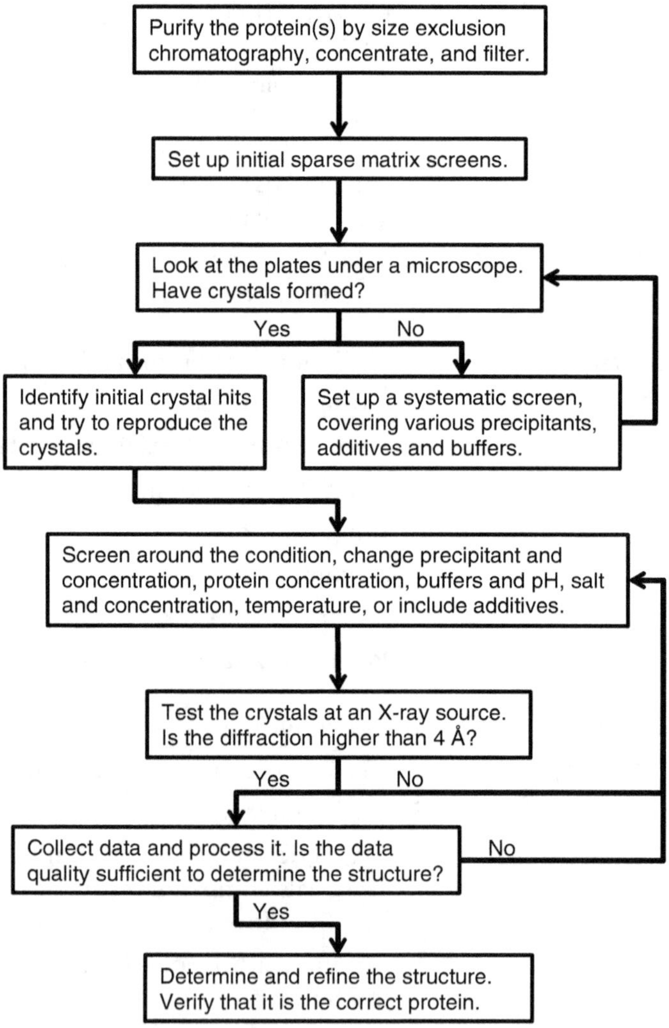

Fig. 1 Schematic over the crystallization pathway

1. Run size exclusion chromatography to get a homogenous sample in a known buffer composition, for instance TBS (*see* **Note 6**).
2. Concentrate the superantigen using a 10,000 molecular weight cut-off (MWCO) concentration tube to 5–10 mg/ml.
3. Determine if the complex can be purified by size exclusion, for instance by running an analytical size exclusion run if the dissociation constant is not known.
4. If the affinity is sufficient for the complex to keep together during the size exclusion run, mix the proteins in an equimolar ratio and purify the complex in one size exclusion run (*see* **Note 7**). If the affinity is too low, purify the superantigen and the receptor/s separately by size exclusion, using the same buffer for all proteins, for instance TBS.
5. If the complex was purified together, concentrate to 5–10 mg/ml and proceed to **step 7**. If the proteins were purified separately, concentrate each protein to 5–10 mg/ml. The superantigen is usually stable at high concentrations and can be concentrated to above 20 mg/ml, which will permit the potentially more unstable proteins to have lower concentrations.
6. Calculate the amount of protein solution needed to set up the desired number of plates with the chosen drop size, including approximately 10 % extra that will be lost due to pipetting. Mix the proteins in an equimolar ratio and dilute with the buffer used in **step 4** to a final concentration of 5–10 mg/ml for the complex.
7. Filter the protein or protein complex through a 0.22 μm spin filter and keep on ice or at 4 °C. Do not purify the protein/s more than a couple of days ahead of doing crystallization trials. If there is an excess of protein or complex and if it is stable at −80 °C, it can be aliquoted and flash frozen in liquid nitrogen and stored at −80 °C.

3.2 Initial Crystal Screening

1. Start with commercial sparse matrix screens with a high amount of polyethylene glycol-based conditions. To save time performing setups, use sitting-drop plates or a crystallization robot if available. If none of the proteins are temperature sensitive, initial screening is carried out at 18–20 °C.
2. Dispense 0.5 ml of each condition in the crystallization screen in the reservoirs of 24-well sitting-drop plates. Let the plates equilibrate at 18–20 °C.
3. Pipet 0.7 μl protein drops on the pedestals of a 24-well sitting drop plate, doing 12 drops at a time (*see* **Note 8**). Without mixing, pipet 0.7 μl reservoir solution on top of each protein drop. When all 12 drops are done, seal them with tape and proceed to the other half of the plate.

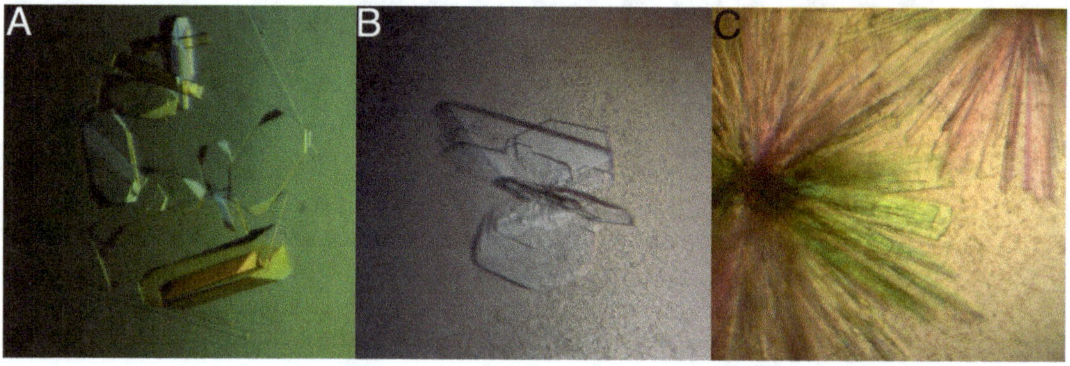

Fig. 2 Examples of superantigen and superantigen–receptor crystals. (**a**) Single crystals of staphylococcal enterotoxin H (SEH), grown in 2-methyl-2,4-pentanediol. (**b**) SEH in complex with a T-cell receptor crystallized as plate clusters in PEG 5000 MME. (**c**) Crystal needle cluster of the ternary complex of SEH, a T-cell receptor, and major histocompatibility complex class II grew in PEG 5000 MME [14]

4. Directly after the plate has been set up, inspect it using a direct light microscope. If about 50 % of the drops are clear and the other half has precipitated, the protein concentration is about right.
5. Incubate the plates in a crystal plate incubator or in a temperature-controlled room. Avoid moving the plate more than necessary and handle with care.
6. Inspect the plates regularly afterwards, for instance at 2 days, 1 week, and 3 weeks. Crystals may appear during a wide time-span, from less than a day to over a year later and can have a number of different appearances (Fig. 2). Other things, such as phase separation or sea urchin-like spheres are also promising signs that the condition is not far from yielding crystals.
7. If no crystals or crystal leads develop, or if there is an abundance of protein, additional screens could be tried, for example CrystalScreen 1 and 2 (Hampton Research), PEG/Ion 1 and 2 (Hampton Research), or Structure Screen 1 and 2 (Molecular Dimensions).

3.3 Further Screening

1. If sparse matrix screens yield no crystals, try systematic screening, by either setting up commercial grid screens or mixing custom systematic screens. For the latter, do 5–10 ml stocks if the screen is to be tried for multiple superantigens or superantigen–receptor complexes. For 10 ml reservoir solution, take 2 ml 0.5 M buffer solution, 1 ml 1 M NaCl solution, and $x/5$ ml, of the 50 % PEG solution where x is the desired final PEG concentration. For concentrations above 35 % PEG, use 100 % PEG solutions and take $x/10$ ml, where x denotes the final PEG concentration. Adjust the volume of each condition to 10 ml. An example screen is provided in Table 1.

Table 1
Suggestion of a systematic PEG-based screen

Plate, reservoirs	Precipitant		Buffer		Salt	
1, A1–D1	25–40 % (v/v)	PEG 200	0.1 M	Sodium acetate pH 4.5	0.1 M	NaCl
1, A2–D2	30–45 % (v/v)	PEG 200	0.1 M	Sodium cacodylate pH 6.5	0.1 M	NaCl
1, A3–D3	25–40 % (v/v)	PEG 350 MME	0.1 M	MES pH 5.5	0.1 M	NaCl
1, A4–D4	30–45 % (v/v)	PEG 350 MME	0.1 M	HEPES pH 7.5	0.1 M	NaCl
1, A5–D5	25–40 % (v/v)	PEG 300	0.1 M	BIS-TRIS PH 6.5	0.1 M	NaCl
1, A6–D6	30–45 % (v/v)	PEG 300	0.1 M	NH_4Cl pH 8.5	0.1 M	NaCl
2, A1–D1	20–35 % (v/v)	PEG 400	0.1 M	Sodium citrate pH 5.0	0.1 M	NaCl
2, A2–D2	25–40 % (v/v)	PEG 400	0.1 M	Imidazole pH 7.0	0.1 M	NaCl
2, A3–D3	20–35 % (v/v)	PEG 550 MME	0.1 M	Sodium phosphate pH 6.0	0.1 M	NaCl
2, A4–D4	25–40 % (v/v)	PEG 550 MME	0.1 M	Tris pH 8.0	0.1 M	NaCl
2, A5–D5	20–35 % (v/v)	PEG 600	0.1 M	PIPES pH 7.0	0.1 M	NaCl
2, A6–D6	25–40 % (v/v)	PEG 600	0.1 M	Glycine pH 9.0	0.1 M	NaCl
3, A1–D1	15–30 % (w/v)	PEG 1000	0.1 M	Sodium acetate pH 4.5	0.1 M	NaCl
3, A2–D2	20–35 % (w/v)	PEG 1000	0.1 M	Sodium cacodylate pH 6.5	0.1 M	NaCl
3, A3–D3	15–30 % (w/v)	PEG 1500	0.1 M	MES pH 5.5	0.1 M	NaCl
3, A4–D4	20–35 % (w/v)	PEG 1500	0.1 M	HEPES pH 7.5	0.1 M	NaCl
3, A5–D5	15–30 % (w/v)	PEG 2000 MME	0.1 M	BIS-TRIS PH 6.5	0.1 M	NaCl
3, A6–D6	20–35 % (w/v)	PEG 2000 MME	0.1 M	NH_4Cl pH 8.5	0.1 M	NaCl
4, A1–D1	15–30 % (w/v)	PEG 2000	0.1 M	Sodium citrate pH 5.0	0.1 M	NaCl
4, A2–D2	20–35 % (w/v)	PEG 2000	0.1 M	Imidazole pH 7.0	0.1 M	NaCl
4, A3–D3	10–25 % (w/v)	PEG 3000	0.1 M	Sodium phosphate pH 6.0	0.1 M	NaCl
4, A4–D4	15–30 % (w/v)	PEG 3000	0.1 M	Tris pH 8.0	0.1 M	NaCl
4, A5–D5	10–25 % (w/v)	PEG 3350	0.1 M	PIPES pH 7.0	0.1 M	NaCl
4, A6–D6	15–30 % (w/v)	PEG 3350	0.1 M	Glycine pH 9.0	0.1 M	NaCl
5, A1–D1	10–25 % (w/v)	PEG 4000	0.1 M	Sodium acetate pH 4.5	0.1 M	NaCl
5, A2–D2	15–30 % (w/v)	PEG 4000	0.1 M	Sodium cacodylate pH 6.5	0.1 M	NaCl
5, A3–D3	10–25 % (w/v)	PEG 5000 MME	0.1 M	MES pH 5.5	0.1 M	NaCl
5, A4–D4	15–30 % (w/v)	PEG 5000 MME	0.1 M	HEPES pH 7.5	0.1 M	NaCl
5, A5–D5	5–20 % (w/v)	PEG 6000	0.1 M	BIS-TRIS PH 6.5	0.1 M	NaCl
5, A6–D6	10–25 % (w/v)	PEG 6000	0.1 M	NH_4Cl pH 8.5	0.1 M	NaCl

(continued)

Table 1 (continued)

Plate, reservoirs	Precipitant		Buffer		Salt	
6, A1–D1	5–20 % (w/v)	PEG 8000	0.1 M	Sodium citrate pH 5.0	0.1 M	NaCl
6, A2–D2	10–25 % (w/v)	PEG 8000	0.1 M	Imidazole pH 7.0	0.1 M	NaCl
6, A3–D3	4–16 % (w/v)	PEG 10000	0.1 M	Sodium phosphate pH 6.0	0.1 M	NaCl
6, A4–D4	8–20 % (w/v)	PEG 10000	0.1 M	Tris pH 8.0	0.1 M	NaCl
6, A5–D5	3–12 % (w/v)	PEG 20000	0.1 M	PIPES pH 7.0	0.1 M	NaCl
6, A6–D6	6–15 % (w/v)	PEG 20000	0.1 M	Glycine pH 9.0	0.1 M	NaCl

2. Set up this screen in the same manner as described in Subheading 3.2. Inspect the plate using a direct light microscope at regular intervals, for example 3 days, 1 week, and 3 weeks.
3. If no crystals are obtained, try different commercial grid screens, change salt of the custom-made screen to for example $MgCl_2$, $CaCl_2$, or KCl. Change the protein buffer from TBS to another buffer. A drop with a promising precipitate (white and with a regular pattern) could be chosen and optimized around, as described in Subheading 3.5. If crystals are still not formed, molecular biology approaches can be used to remove potential disordered loops or free cysteines.

3.4 Initial Hit Identification

1. Once crystals start to form in a condition, putative salt crystals must be ruled out. Inspect the crystals under the microscope with a polarizing filter. Both protein and salt crystals are birefringent, but protein crystals less so, and will display fewer colors than salt crystals. This is by no means a safe way of distinguishing protein from salt, but may give a hint. In addition, if the aim is to get a protein–protein complex, it is possible that only one of the proteins has crystallized, or if the complex has. The most certain way to distinguish between the possible crystals is to collect diffraction data. If this is not practically possible, a 24-well test plate can be set up.
2. Try to reproduce the crystals in the same condition. For example, if the aim is to crystallize a superantigen in complex with a T-cell receptor and crystals appear in 20 % (w/v) PEG 4000, 0.1 M HEPES pH 7.5, 0.2 M $MgCl_2$, at a total protein concentration of 10 mg/ml, set up a plate with row 4 set at 20 % (w/v) PEG 4000 screening with 2–3 % steps. For instance, 3 % steps would yield a plate with column 1 at 11 %, 2 at 14 %, 3 at 17 %, 4 at 20 %, 5 at 23 %, and 6 at 26 %.

3. Pipet 0.7 μl of buffer only or protein solution onto the pedestals of the plate. Pipet 0.7 μl of the respective reservoir solution onto each drop and seal with clear tape. Set up row A with buffer without any protein, row B with SAg at 3.5 mg/ml (given that the SAg–TCR protein concentration is 10 mg/ml and the molecular weights of the SAg and TCR are 27 kDa and 50 kDa respectively), row C with TCR at 6.5 mg/ml, and row D with SAg-TCR at 10 mg/ml (*see* **Note 9**).

3.5 Crystal Hit Optimization

There are a number of different ways of optimizing crystals, and all optimization trials do not lead to improved crystals. Thus, the points listed below do not have to be tried in the exact order, and some may work while others do not.

1. Once crystals are yielded, it may be beneficial to try to reproduce the crystals in hanging-drop plates instead of sitting drop plates, since it may be more convenient to mount crystals from hanging-drop plates. Start with the original condition and screen around the PEG concentration. For instance, if the condition is 20 % (w/v) PEG 3350, 0.1 M BIS-TRIS pH 6.5 and 0.1 M NaCl, set up a plate from 11 to 26 % PEG 3350, in 3 % steps in columns 1–6 and 0.1 M BIS-TRIS pH 6.0–7.5 in 0.5 steps. Do this by pipetting 110–260 μl 50 % (w/v) PEG 3350 in each well, doing 30 μl increments between columns 1 and 6. Use four 0.5 M BIS-TRIS stocks at pH 6.0, 6.5, 7.0, and 7.5 and pipet 100 μl 0.5 M BIS-TRIS stock per well and use a different pH for each row A–D. Pipet 50 μl 1 M NaCl in each well and make up the volume of each to 500 μl with ultrapure water.
2. Cover the rim of each well with oil. Line up six siliconized cover slips (18 mm in diameter) and pipet 1 μl protein drops on top of each cover. Directly afterwards, pipet 1 μl reservoir solution on top of each protein drop, without mixing. When all six drops are done, use a pair of tweezers to flip the cover slip over the reservoir. Take care to check that it is sealed completely by the oil. When all six are done, proceed to the next row.
3. Set up an additive screen. Use the initial reservoir solution and prepare 50–100 ml of this condition. Pipet 500 μl in all wells of a 24-well plate and cover the rims of all wells with oil. Pipet 6×4 μl drops of precipitant solution on saran wrap and add 1 μl of each condition in the additive screen to each drop and mix. Then, pipet 1 μl protein drops on six cover slips and subsequently pipet 1 μl of each reservoir+additive mix onto each drop, turn the cover slips over and seal the wells. For the volatile additives, add 900 μl reservoir solution and 100 μl additive to the well and add 1 μl of this to the protein drop.
4. Try different salts, such as $MgCl_2$, $CaCl_2$, and KCl.

5. Vary drop size from the usual 1 μl protein and 1 μl reservoir to 2 μl + 1 μl protein + reservoir, 1 μl + 2 μl protein + reservoir, and 2 μl + 2 μl protein + reservoir. For hanging drops, up to 3 μl + 3 μl can be done. Two or three hanging drops can be made on each cover slip, in order to decrease the number of plates in total. When doing multiple drops, it may be difficult to do six cover slips at a time without the drops drying out. Then, do three by three or less.
6. Set up drops at different temperatures, such as 4 °C.

3.6 Crystal Mounting

1. Once crystals of at least 0.03 mm have been obtained, it is possible to mount them for screening using X-ray diffraction. Prepare cryo-protectant solution by mixing either 800 μl of the commercial screen condition with 200 μl 98 % glycerol or 100 % PEG 400 or prepare 1 ml of the identical reservoir solution using the stock solutions, but adding 200 μl of 100 % (v/v) glycerol or PEG 400 (*see* **Note 10**). Choose the appropriate loop sizes and glue the loop pins to magnetic cryo caps. Prepare some different sizes and glue some extra to avoid running out of them. Fill a canister with liquid nitrogen to the brim and freeze a cryo cane with a cryo vial (*see* **Note 11**).
2. Open the reservoir by either cutting the tape with a scalpel or turn the cover slip over with a pair of tweezers (*see* **Note 12**). Remove any skin from the drop and pipet a 1 μl drop of cryo-protectant either directly on the drop or beside the drop. For sitting drops, the easiest is to add the cryo-protectant on top of the drop, whereas for sitting drops, there are some benefits with adding the cryo drop beside the crystal drop. For instance, different cryo-protectants can be tried, and the drop can be sealed again afterwards, without risking that the crystals are destroyed.
3. Mount the crystal in the cryo loop and, if a separate cryo drop has been made, transfer the crystal to this drop before freezing it in liquid nitrogen. The freezing process should be carried out as quickly as possible, to avoid ice formation. Take care to mount only one separate crystal in each loop, if possible. Break apart crystals that have grown together, if necessary.
4. Transfer the crystal loop to the vial, taking care not to raise the frozen crystal above the liquid nitrogen surface once it has been frozen, since the crystal will thaw rapidly and possibly be destroyed by this. Once the loop is transferred to the vial, add an empty vial to the cane and mount a new crystal in a new loop.
5. If the drop seems to dry during the process of mounting crystals, add 1 μl reservoir solution on the drop. After all the desired crystals have been recovered from the drop, seal it again in order to preserve any remaining crystals.
6. Store the crystals in a liquid nitrogen dewar until it is possible to screen them at an X-ray source (*see* **Note 13**).

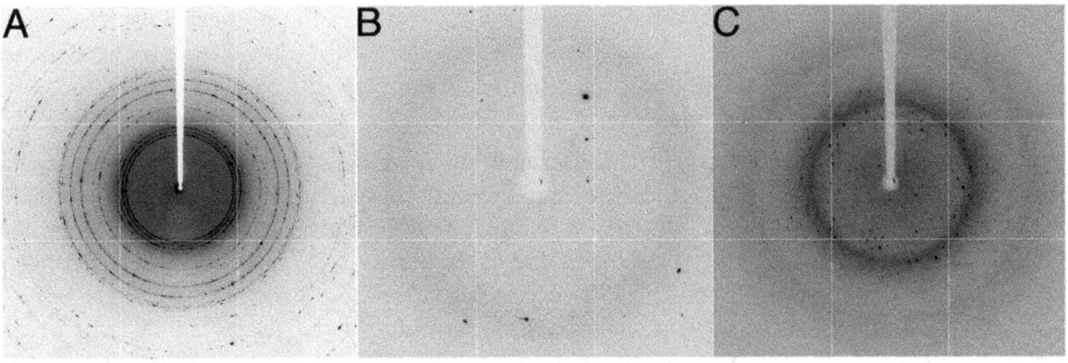

Fig. 3 Examples of diffraction patterns. (**a**) Ice rings from ice crystals formed in the absence of cryo-protectant. (**b**) Diffraction from a salt crystal. (**c**) Diffraction pattern obtained from a staphylococcal enterotoxin D crystal

3.7 Diffraction Data Collection

1. Mount the loop with the frozen crystal in the cryo stream at an in-house X-ray source or synchrotron beamline, and center it in the X-ray beam.
2. Take test shots at a fixed wavelength, for example 1 Å, at 0°, 45° and 90°, with 1° oscillation and look at the diffraction pattern. If the diffraction pattern displays several rings, there are ice crystals present due to freezing issues and the cryo needs optimization (Fig. 3a), and if there are a few large diffraction spots, this corresponds to salt crystals (Fig. 3b). Protein crystals give rise to diffraction patterns with several small spots (Fig. 3c).
3. If the crystal diffract well (to higher than 4 Å), collect data directly and do not take the crystal down (*see* **Note 14**). Even though the space group may allow for collection of less than 180°, collect 180° anyway, since the space group may be wrong or the crystal twinned so that more data are needed. If the suggested space group is P1, rotate the crystal 30° around the κ angle and collect an additional 60° if this is possible. If the resolution is better than 1.5 Å, start with collecting a low resolution data set to 3 Å before collecting the high resolution data set to the maximum diffracting power of the crystal.
4. If there are other data sets of the same types of crystals previously, the angle may be reduced from 180°, depending on the space group. Try to collect a complete data set rather than to maximize resolution.

3.8 Structure Determination

In order to determine the structure, several computer programs are necessary. The software mentioned in **steps 1–5** are only examples, and other software capable of performing the tasks are available and can be used instead.

1. Index and integrate the reflections using XDS [21], or autoPROC [22].
2. Scale and merge the data using aimless [23, 24] either within CCP4 [25] or autoPROC [22], and decide on a resolution

cut-off, for instance at I/σ at 2 in the highest resolution shell, and completeness in this shell not below 80 %. If the data are integrated to a higher resolution, go back to **step 1** and limit the high resolution limit.

3. Inspect the data, taking note of any systematic absences that may indicate that the chosen space group is wrong. Calculate the Matthew's coefficient which, given the molecular weight of the proteins in the asymmetric unit, gives the solvent content [26, 27]. Since crystals of soluble proteins contain 40–60 % solvent, it is possible to judge if the molecular weight of the protein or protein complex may fit in the asymmetric unit, or if there are multiple copies present.
4. Solve the structure using molecular replacement. For superantigens that have been determined previously, use the structure as a search model. If not, do a sequence alignment and chose the most similar superantigen. Substitute any differing amino acids to alanine and remove any differing loops. The same can be applied for TCR and/or MHC class II. Use separate chains as input models in Phaser [28], and allow the program to test all alternative space groups. If multiple protein compositions are possible, for instance if the Matthew's coefficient suggests that the contents of the asymmetric unit is 75 kDa, this could be both three superantigen molecules (3 times 25 kDa), or a superantigen–TCR complex (25 kDa plus 50 kDa), both possibilities should be tried in Phaser, and judged by Z-score and by evaluating the electron density maps in coot.
5. Refine the structure using refmac5 [29] or autoBUSTER [30], with rounds of manual model building in coot [31] (*see* **Note 15**). Aim for a R-value of less than 10 times the resolution (in percentages) and a discrepancy of the R-value and R-free of less than 5 %. The Ramachandran outliers should be kept as close to 0 as possible. RMS deviations from ideal bond lengths should be <0.02 Å and ideal bond angles <2°.

4 Notes

1. Make sure that the buffer is room tempered, since the pH of Tris buffer is temperature dependent.
2. If it is not possible to filter through a 0.22 μm filter, use a 0.45 μm filter instead.
3. MES buffer decomposes relatively fast and turns yellow. Store at 4 °C and discard if it is yellow.
4. Sodium cacodylate and imidazole are toxic and can be replaced or omitted from the screen if desired. However, some commercial screens contain these buffers.

5. It is important to use cryo caps and cryo vials compatible with the goniometer and sample changer for the particular X-ray source, in order to avoid damage that may be caused by incompatible equipment. If possible, keep only one type of equipment in order to avoid mistakes.
6. Always wear gloves and lab coat when working with superantigens, and eye protection if splashes may occur, since they are toxic and any contact should be avoided. Strong bases irreversibly denature superantigens. Therefore, keep a bottle of 1 M sodium hydroxide close when working with SAgs, and use for decontamination of any spill.
7. If one protein is easier to obtain, it could be added in excess in order to loose less of the more precious protein. Any monomeric proteins will be separated from the complex on the size exclusion column.
8. This step should be carried out as quickly as possible, to prevent drops from drying. When pipetting the protein solution, it is not necessary to change tips in between drops. Also avoid creating bubbles during this step, for example by dispensing only to the first stop of the pipette and not to the second stop. It is more important to be consistent than to get a drop size of exactly 1.4 μl.
9. Care should be taken using this method to try to distinguish if the crystal contains multiple proteins, since there are instances where the complex crystallizes in the same condition as one of the single proteins.
10. Some conditions may not require any cryo-protectant, since they may be providing adequate cryo-protection by themselves. As a rule of thumb, 30 % of the low molecular weight PEGs (such as PEG 1000) is sufficient. In addition, high concentrations of for instance sucrose, glycerol, 2-propanol or 2-methyl-2,4-pentanediol (MPD) are also acting as cryo-protectants at high concentrations. If in doubt, test the reservoir solution by dipping a cryo loop in it and mount it in the cryo stream and collect a single diffraction image. If there are no ice rings, additional cryo-protectants are not needed.
11. Note that the canister needs to be filled to the brim, as a gap between the brim and the level of liquid nitrogen will result in a slower freezing process of the crystal and may result in ice formation.
12. In order to minimize the time the drop is exposed to air, prepare everything before and have all necessary equipment available, otherwise, the drop may dry out before any crystals are recovered.
13. If it is possible to transport the unfrozen crystals to the beamline, they can be frozen directly in the cryostream, which may result in better data quality compared to if the crystals are frozen prior to transportation to the beamline.

14. There are always risks associated with dismounting the crystal, such as manual handling errors or sample changer malfunctioning resulting in loss of the crystal. In addition, just the temperature changes associated with dismounting the crystal may result in loss of diffraction quality. Therefore, if it is the first crystal that diffracts to higher than 4 Å, collect data directly.
15. Always use the output mtz-file from aimless as the file that refmac or autoBUSTER refines against. Do not use the output mtz-file from the previous run, since this will introduce model bias.

Acknowledgements

This work was funded by the Swedish Research Council (VR-2011-2891), the Cancer Foundation, the Konsul Thure Carlsson's Foundation, the Olle Engkvist Byggmästare Foundation, and the Åke Wiberg Foundation.

References

1. Fraser JD, Proft T (2008) The bacterial superantigen and superantigen-like proteins. Immunol Rev 225:226–243. doi:10.1111/j.1600-065X.2008.00681.x
2. Spaulding AR, Salgado-Pabon W, Kohler PL, Horswill AR, Leung DY, Schlievert PM (2013) Staphylococcal and streptococcal superantigen exotoxins. Clin Microbiol Rev 26:422–447. doi:10.1128/CMR.00104-12, 26/3/422 [pii]
3. Swaminathan S, Furey W, Pletcher J, Sax M (1992) Crystal structure of staphylococcal enterotoxin B, a superantigen. Nature 359:801–806. doi:10.1038/359801a0
4. Jardetzky TS, Brown JH, Gorga JC, Stern LJ, Urban RG, Chi YI, Stauffacher C, Strominger JL, Wiley DC (1994) Three-dimensional structure of a human class II histocompatibility molecule complexed with superantigen. Nature 368:711–718. doi:10.1038/368711a0
5. Fields BA, Malchiodi EL, Li H, Ysern X, Stauffacher CV, Schlievert PM, Karjalainen K, Mariuzza RA (1996) Crystal structure of a T-cell receptor beta-chain complexed with a superantigen. Nature 384:188–192. doi:10.1038/384188a0
6. Buonpane RA, Moza B, Sundberg EJ, Kranz DM (2005) Characterization of T cell receptors engineered for high affinity against toxic shock syndrome toxin-1. J Mol Biol 353:308–321. doi:10.1016/j.jmb.2005.08.041, S0022-2836(05)00978-2 [pii]
7. Buonpane RA, Churchill HR, Moza B, Sundberg EJ, Peterson ML, Schlievert PM, Kranz DM (2007) Neutralization of staphylococcal enterotoxin B by soluble, high-affinity receptor antagonists. Nat Med 13:725–729. doi:10.1038/nm1584, nm1584 [pii]
8. Wang N, Mattis DM, Sundberg EJ, Schlievert PM, Kranz DM (2010) A single, engineered protein therapeutic agent neutralizes exotoxins from both *Staphylococcus aureus* and *Streptococcus pyogenes*. Clin Vaccine Immunol 17:1781–1789. doi:10.1128/CVI.00277-10, CVI.00277-10 [pii]
9. Woody MA, Krakauer T, Stiles BG (1997) Staphylococcal enterotoxin B mutants (N23K and F44S): biological effects and vaccine potential in a mouse model. Vaccine 15:133–139, doi:S0264-410X(96)00166-1 [pii]
10. McCormick JK, Tripp TJ, Olmsted SB, Matsuka YV, Gahr PJ, Ohlendorf DH, Schlievert PM (2000) Development of streptococcal pyrogenic exotoxin C vaccine toxoids that are protective in the rabbit model of toxic shock syndrome. J Immunol 165:2306–2312, doi:ji_v165n4p2306 [pii]
11. Krupka HI, Segelke BW, Ulrich RG, Ringhofer S, Knapp M, Rupp B (2002) Structural basis for abrogated binding between staphylococcal enterotoxin A superantigen vaccine and MHC-IIalpha. Protein Sci 11:642–651. doi:10.1110/ps.39702
12. Bolin DR, Swain AL, Sarabu R, Berthel SJ, Gillespie P, Huby NJ, Makofske R, Orzechowski L, Perrotta A, Toth K, Cooper JP, Jiang N, Falcioni F, Campbell R, Cox D, Gaizband D,

Belunis CJ, Vidovic D, Ito K, Crowther R, Kammlott U, Zhang X, Palermo R, Weber D, Guenot J, Nagy Z, Olson GL (2000) Peptide and peptide mimetic inhibitors of antigen presentation by HLA-DR class II MHC molecules. Design, structure-activity relationships, and X-ray crystal structures. J Med Chem 43:2135–2148

13. Sundberg EJ, Sawicki MW, Southwood S, Andersen PS, Sette A, Mariuzza RA (2002) Minor structural changes in a mutated human melanoma antigen correspond to dramatically enhanced stimulation of a CD4+ tumor-infiltrating lymphocyte line. J Mol Biol 319:449–461. doi:10.1016/S0022-2836(02)00370-4
14. Saline M, Rodstrom KE, Fischer G, Orekhov VY, Karlsson BG, Lindkvist-Petersson K (2010) The structure of superantigen complexed with TCR and MHC reveals novel insights into superantigenic T cell activation. Nat Commun 1:119. doi:10.1038/ncomms1117
15. Rodstrom KE, Elbing K, Lindkvist-Petersson K (2014) Structure of the superantigen staphylococcal enterotoxin B in complex with TCR and peptide-MHC demonstrates absence of TCR-peptide contacts. J Immunol 193:1998–2004. doi:10.4049/jimmunol.1401268, jimmunol.1401268 [pii]
16. Sundberg EJ, Li H, Llera AS, McCormick JK, Tormo J, Schlievert PM, Karjalainen K, Mariuzza RA (2002) Structures of two streptococcal superantigens bound to TCR beta chains reveal diversity in the architecture of T cell signaling complexes. Structure 10:687–699
17. Li Y, Li H, Dimasi N, McCormick JK, Martin R, Schuck P, Schlievert PM, Mariuzza RA (2001) Crystal structure of a superantigen bound to the high-affinity, zinc-dependent site on MHC class II. Immunity 14:93–104
18. Schad EM, Zaitseva I, Zaitsev VN, Dohlsten M, Kalland T, Schlievert PM, Ohlendorf DH, Svensson LA (1995) Crystal structure of the superantigen staphylococcal enterotoxin type A. EMBO J 14:3292–3301
19. Papageorgiou AC, Brehm RD, Leonidas DD, Tranter HS, Acharya KR (1996) The refined crystal structure of toxic shock syndrome toxin-1 at 2.07 A resolution. J Mol Biol 260:553–569
20. Roussel A, Anderson BF, Baker HM, Fraser JD, Baker EN (1997) Crystal structure of the streptococcal superantigen SPE-C: dimerization and zinc binding suggest a novel mode of interaction with MHC class II molecules. Nat Struct Biol 4:635–643
21. Kabsch W (1993) Automatic processing of rotation diffraction data from crystals of initially unknown symmetry and cell constants. J Appl Crystallogr 26:795–800
22. Vonrhein C, Flensburg C, Keller P, Sharff A, Smart O, Paciorek W, Womack T, Bricogne G (2011) Data processing and analysis with the autoPROC toolbox. Acta Crystallogr Sect D Biol Crystallogr 67:293–302. doi:10.1107/S0907444911007773, S0907444911007773 [pii]
23. Evans P (2006) Scaling and assessment of data quality. Acta Crystallogr Sect D Biol Crystallogr 62:72–82. doi:10.1107/S0907444905036693, S0907444905036693 [pii]
24. Evans PR (2011) An introduction to data reduction: space-group determination, scaling and intensity statistics. Acta Crystallogr Sect D Biol Crystallogr 67:282–292. doi:10.1107/S090744491003982X, S090744491003982X [pii]
25. Collaborative Computational Project, Number 4 (1994) The CCP4 suite: programs for protein crystallography. Acta Crystallogr Sect D Biol Crystallogr 50:760–763. doi:10.1107/S0907444994003112
26. Matthews BW (1968) Solvent content of protein crystals. J Mol Biol 33:491–497
27. Kantardjieff KA, Rupp B (2003) Matthews coefficient probabilities: improved estimates for unit cell contents of proteins, DNA, and protein-nucleic acid complex crystals. Protein Sci 12:1865–1871. doi:10.1110/ps.0350503
28. McCoy AJ, Grosse-Kunstleve RW, Adams PD, Winn MD, Storoni LC, Read RJ (2007) Phaser crystallographic software. J Appl Crystallogr 40: 658–674. doi:10.1107/S0021889807021206
29. Murshudov GN, Vagin AA, Dodson EJ (1997) Refinement of macromolecular structures by the maximum-likelihood method. Acta Crystallogr Sect D Biol Crystallogr 53:240–255. doi:10.1107/S0907444996012255
30. Bricogne G, Blanc E, Brandl M, Flensburg C, Keller P, Paciorek W, Roversi P, Sharff A, Smart OS, Vonrhein C, Womack TO (2011) BUSTER version 2.10.1. Global Phasing Ltd, Cambridge
31. Emsley P, Cowtan K (2004) Coot: model-building tools for molecular graphics. Acta Crystallogr Sect D Biol Crystallogr 60:2126–2132. doi:10.1107/S0907444904019158

Chapter 5

Lipopolysaccharide-Induced Toxic Shock Syndrome in Rabbits

Christopher S. Stach and Patrick M. Schlievert

Abstract

Enhancement of susceptibility to lipopolysaccharide (LPS; endotoxin) is a defining characteristic of *Staphylococcus aureus* superantigens. At the time of this publication, there are 24 identified staphylococcal superantigens (SAgs), some of which have yet to be fully characterized. Testing the capacity of superantigens to potentiate LPS sensitivity is essential to characterize the role of these proteins in disease development. Here we describe how to perform studies of the enhancement of LPS-induced toxic shock syndrome in rabbits. This protocol also provides information on a second important activity of superantigens: the production of fever.

Key words Superantigens, *Staphylococcus aureus*, Endotoxin enhancement, LPS-induced toxic shock syndrome

1 Introduction

Superantigens (Sags) are a unique family of exotoxins produced by gram-positive cocci that interact directly with the immune system and the epithelium/endothelium during infection [1, 2]. The protein family in *S. aureus* is composed of 23 distinct members designated as staphylococcal enterotoxins (SE) or enterotoxin-like (SEl) toxins and toxic shock syndrome toxin-1 (TSST-1) [2]. The term superantigen refers to the ability of the proteins to induce extensive T-cell activation and proliferation [3]. For comparison, a classical antigen will stimulate 0.01 % T-cells (1 in 10,000 T-cells) whereas SAgs can stimulate up to 50 % of T-cells [1]. SAgs share common phenotypic traits such as pyrogenicity, T-cell mitogenicity, and enhancement of lipopolysaccharide (LPS; endotoxin) shock [3–5].

SAgs are secreted proteins that range in size from 19 to 30 kDa and are resistant to denaturation by heat, proteolysis, and chemical treatments [2]. They are composed of a carboxy-terminal beta-grasp domain and an amino-terminal oligosaccharide/oligonucleotide

Amanda J. Brosnahan (ed.), *Superantigens: Methods and Protocols*, Methods in Molecular Biology, vol. 1396,
DOI 10.1007/978-1-4939-3344-0_5,

binding (OB) fold [1, 6]. Some SAgs also contain a cysteine loop which is responsible for emetic activity [7]. Recently, a conserved cell binding region known as the dodecapeptide domain has also been identified [1]; this region appears to interact with epithelial and endothelial cells.

The mode of action of SAgs involves crosslinking T-cell receptors (TCRs) and major histocompatibility complex class II molecules (MHCIIs), leading to massive chemokine and cytokine release [6]. Specifically, the proteins bind to the beta-chain variable region of the TCR (Vbeta-TCR) and the alpha and/or beta chain of MHCII on antigen presenting cells, crosslinking the molecules and resulting in T-cell and antigen presenting cell activation, and production of pro-inflammatory chemokines and cytokines [8]. Each SAg displays a unique Vbeta-TCR binding profile [6, 9].

The capacity to sensitize rabbits by up to 10^6-fold to LPS exposure is a hallmark feature of all pyrogenic toxin SAgs. Rabbits are the most sensitive animal model for in vivo analysis of LPS-induced toxic shock syndrome (TSS) because rabbits exhibit similar symptoms and lethality to human TSS [10]. SAgs increase sensitivity to LPS, apparently by: (1) inducing systemic inflammation through synergized production of tumor necrosis factor alpha, and (2) increasing circulating LPS after direct inhibition of liver clearance function [11]. In this chapter we will outline the protocol to determine the capacity of SAgs to enhance LPS-mediated TSS in rabbits. It is important to note that although mice may exhibit up to a 20-fold enhancement of LPS-mediated TSS by SAgs, mice remain highly resistant and have little practical use in these studies.

2 Materials

All solutions are prepared using ultrapure water (sensitivity of 18 MΩ cm at 25 °C) and analytical, pyrogen-free grade reagents. All materials and reagents must be LPS free. Glassware is baked at 160 °C for 3 h to carbonize LPS contamination.

1. Phosphate-buffered saline (PBS): 5 mM Na_2HPO_4/NaH_2PO_4 buffer, 150 mM NaCl, at pH 7.2, sterilized by autoclaving.
2. LPS (endotoxin): LPS was prepared by the method described by Westphal et al. [12]. All of the experiments performed in this chapter use LPS purified from *Salmonella enterica* serovar Typhimurium dissolved in PBS.
3. Superantigens: All superantigens are dissolved in sterile PBS for in vivo analysis. Superantigens were purified from *S. aureus* strains by preparative isoeletric focusing as described in Chapter 2 in this volume. Superantigens may be purified from *Escherichia coli* using over-expression constructs, but LPS contamination must be removed before use in this assay.

4. Pyrogen-free syringes and needles: 25G×7/8 in. needles are used for intravenous injection, 27G×½ in. needles are used for anesthesia.
5. Surgical anesthesia: Ketamine (Ketaject; Phoenix, St. Joseph, MO) at 100 mg/mL; Xylazine (Anased; Lloyd, Shenandoah, IA) at 20 mg/mL.
6. Euthanasia solution: Euthasol (Virbac AH, Inc., Fort Worth, TX). Per mL: 390 mg pentobarbital sodium (barbituric acid derivative) and 50 mg phenytoin sodium.
7. Rabbits (*see* **Note 1**): New Zealand white rabbits, mixed sex, 1.5–2.0 kg each. We obtain our rabbits from Bakkom Rabbitry (Red Wing, MN).

3 Methods

The methods described in this chapter involve the use of rabbits and as such all protocols must be approved by relevant governing bodies prior to being performed. Laboratory personnel should have relevant training in animal welfare and experimentation prior to starting any protocols involving rabbits.

The LPS enhancement model of TSS will yield information on both the pyrogenicity and the LPS enhancement effects due to SAgs. Pyrogenicity is defined as an average increase in body temperature of 0.5 °C over a 4 h period of time for three rabbits per group. Positive controls for the assay can be any of the highly characterized SAgs such as TSST-1, SEC, or SEB, which are known to have pyrogenic activity and to enhance LPS toxicity in rabbits. Negative controls are doses of PBS equal in volume to the SAg doses. All intravenous injections are done through the marginal ear veins.

1. Acclimate rabbits to the experimental environment.
2. Take rectal temperatures to determine a baseline temperature for each rabbit (*see* **Note 2**).
3. Prepare injection site by removing hair through careful shaving or use of a chemical depilatory agent to expose marginal ear vein (*see* **Note 3**).
4. Prepare SAgs at a concentration of 5 μg/mL in sterile pyrogen-free PBS.
5. Inject SAgs through the marginal ear vein at a dosage of 5 μg/kg for each rabbit. An equal volume of only PBS is injected into the control cohort.
6. Monitor temperatures every hour over a period of 4 h to determine pyrogenicity. SAgs cause a fairly linear rise in rabbit body temperature, peaking at 4 h (*see* **Note 4**).

7. LPS enhancement is done at 4 h post SAg injection (*see* **Note 5**). Prepare LPS at a concentration *up to* 10 μg/mL (1/50 LD_{50} of LPS) in sterile pyrogen-free PBS (*see* **Note 6**). Lower LPS doses may be used depending on the purpose of experimentation (*see* **Note 7**).
8. Inject LPS through the marginal ear vein of all rabbits.
9. Monitor rabbits closely for development of lethality due to TSS symptoms. Symptoms include fever with subsequent rapid development of hypothermia indicating shock is developing, diarrhea, labored breathing, absence of escape response, and failure of animals to right themselves (*see* **Note 8**).
10. Mortality, which is defined as death or premature euthanasia of rabbits, is monitored for up to 48 h after LPS injection. Most often, death occurs in 1–2 h post LPS injection, but if doses are low, death may occur up to 48 h.
11. All animals surviving past the 48-h time point must be euthanized. Animals are sedated and subsequently euthanized by intravenous administration of 1 mL of Euthasol solution (*see* **Note 9**).

4 Notes

1. Only healthy rabbits should be used for experiments. Samples of 0.1 mL of blood from each animal should be streaked on blood agar plates prior to use to verify the health of the animals. Rabbits with infections will yield inconsistent results and introduce variability into the assay.
2. Rabbits are immobilized by installation into an approved restraining device. Temperatures are taken rectally using approved and standardized thermocouples. Thermocouples are lubricated using petroleum jelly to minimize discomfort to the animal.
3. Marginal ear veins should be exposed on both ears.
4. Generally, temperature will peak at 4 h post SAg injection. The fever response of rabbits given SAgs differs from that of rabbits given LPS. LPS causes fever peaks at both 1 and 3 h post intravenous injection.
5. Four hours post SAg injection has been experimentally shown to be the optimal time to introduce LPS to determine capacity of the SAg to increase LPS sensitivity of rabbits [10].
6. This dose of LPS is 1/50th of the lethal dose of LPS alone as previously determined [10].
7. There is a log:log relationship between SAg pretreatment and LD_{50} of LPS, with a slope of −1. This means that for every log

increase in SAg dose for pretreatment, the dose of LPS may be decreased by a log. There is not LD_{50} of SAgs alone by bolus injection in rabbits. To date, we have used doses of TSST-1 as high as 5000 μg/ml without lethality. The LD_{50} of LPS in rabbits is 500 μg/kg by ear vein injection. It is also important to recognize that rabbits must be pretreated with SAg for at least 2 h and less than 12 h prior to administration of LPS. If SAg and LPS are given at the same time or if LPS is administered ahead of SAg, neither fever nor LPS enhancement will be seen [10].

8. In our approved protocols, animals must be euthanized if they display an absence of normal escape response combined with failure to right themselves. These indications are 100 % predictive of subsequent death.
9. Animals are first anesthetized by subcutaneous injection of ketamine (40–90 mg/kg) and xylazine (3–5 mg/kg) using 27G × ½ in. needles.

References

1. Brosnahan AJ, Schlievert PM (2011) Gram-positive bacterial superantigen outside-in signaling causes toxic shock syndrome. FEBS J 278:4649–4667
2. Kotb M, Fraser JD, American Society for Microbiology (2007) Superantigens: molecular basis for their role in human diseases. ASM Press, Washington, DC
3. Marrack P, Kappler J (1990) The staphylococcal enterotoxins and their relatives. Science 248:705–711
4. Leung DYM, Huber BT, Schlievert PM (1997) Superantigens: molecular biology, immunology, and relevance to human disease. Marcel Dekker, New York
5. Bohach GA, Fast DJ, Nelson RD, Schlievert PM (1990) Staphylococcal and streptococcal pyrogenic toxins involved in toxic shock syndrome and related illnesses. Crit Rev Microbiol 17:251–272
6. Fraser JD, Proft T (2008) The bacterial superantigen and superantigen-like proteins. Immunol Rev 225:226–243
7. Hovde CJ, Marr JC, Hoffmann ML, Hackett SP, Chi YI, Crum KK, Stevens DL, Stauffacher CV, Bohach GA (1994) Investigation of the role of the disulphide bond in the activity and structure of staphylococcal enterotoxin C1. Mol Microbiol 13:897–909
8. Krakauer T (2012) PI3K/Akt/mTOR, a pathway less recognized for staphylococcal superantigen-induced toxicity. Toxins (Basel) 4:1343–1366
9. Choi Y, Lafferty JA, Clements JR, Todd JK, Gelfand EW, Kappler J, Marrack P, Kotzin BL (1990) Selective expansion of T cells expressing V beta 2 in toxic shock syndrome. J Exp Med 172:981–984
10. Schlievert PM (1982) Enhancement of host susceptibility to lethal endotoxin shock by staphylococcal pyrogenic exotoxin type C. Infect Immun 36:123–128
11. Schlievert PM (1983) Alteration of immune function by staphylococcal pyrogenic exotoxin type C: possible role in toxic-shock syndrome. J Infect Dis 147:391–398
12. Westphal O, Luderitz O, Bister F (1952) Uber Die Extraktion Von Bakterien Mit Phenol Wasser. Z Naturforsch B 7:148–155

Chapter 6

Aortic Valve Damage for the Study of Left-Sided, Native Valve Infective Endocarditis in Rabbits

Wilmara Salgado-Pabón and Patrick M. Schlievert

Abstract

Infective endocarditis affects approximately 100,000 individuals in the USA. Medical advances have contributed to the rise of the disease, and no new therapies have emerged in the last 50 years to control the surge of this life-threatening infection. The rabbit vascular physiology and immune response mechanisms are similar to humans. Hence, the rabbit model of infective endocarditis is an excellent research tool with which to address many questions regarding development of endocarditis, for the testing of new therapies, and for the study of the molecular mechanisms used by infectious agents to cause disease. This chapter describes the surgical procedure required to study infective endocarditis in damaged native valves, therefore closely mimicking human disease.

Key words Native valve endocarditis, Carotid catheterization, Rabbits, Aortic valve damage

1 Introduction

Infective endocarditis is an infection of the cardiac endothelium, often on valve tissue [1, 2]. Structural abnormalities that result in turbulence across high- to low-pressure gradients efface surfaces that are then predisposed to platelet and fibrin deposition, leading to what are referred to as sterile vegetations. In the course of bacteremia (usually transient), organisms as *Staphylococcus aureus*, streptococcus viridans group, or enterococci may colonize the sterile vegetations, causing further accumulation of host factors as fibrin, erythrocytes, and platelets. Enlargement of infected vegetations may lead to valve perforation or destruction leading to regurgitation and congestive heart failure. Infective endocarditis vegetations can also fragment sending septic emboli systemically, resulting in strokes and metastatic infections at distal organs as the kidneys, lungs, liver, and spleen.

Two major animal models have been used to study development of infective endocarditis, the rabbit and the rat [3–13]. The most commonly used model is the rabbit, based on its similar

Amanda J. Brosnahan (ed.), *Superantigens: Methods and Protocols*, Methods in Molecular Biology, vol. 1396,
DOI 10.1007/978-1-4939-3344-0_6, © Springer Science+Business Media New York 2016

cardiovascular system to humans. The standard way to establish infective endocarditis is to feed a catheter into the animal's left carotid artery to abut against the aortic valve and cause damage. When rodent models are used, the catheters have to be left in place to provide continued damage to the valve and adjacent cardiac tissue, and to maintain colonization via biofilm formation. In the native valve rabbit model described herein, the catheters remain in place for 2 h and then are removed. In this way, the model duplicates more closely human development of infective endocarditis than rodent models or rabbit models where the catheter is left in place for the duration of the study [9, 14].

2 Materials

Anesthetics and analgesia should be used at the doses specified in the protocol approved by the Institutional Animal Care and Use Committee (IACUC). Autoclave surgical instruments before use. This procedure requires a minimum of two persons: a surgeon and an assistant (*see* **Note 1**).

2.1 Solutions

1. Ketamine: Ketamine HCl, 10 ml vial, 100 mg/ml.
2. Xylazine: Xylazine HCl, 50 ml vial, 100 mg/ml.
3. Buprenorphine: Buprenorphine HCl, 0.3 mg/ml.
4. Syringes: 5 ml syringe, luer-lok tip.
5. Needles: 25 gauge, $^7/_8$ needle (*see* **Note 2**).
6. Betadine surgical scrub (Povidone-iodine, 7.5 %).
7. Sterile phosphate-buffered saline (PBS, 1×): 0.005 M $NaPO_4$, pH 7.2, 0.15 M NaCl.

2.2 Equipment

1. Electrical razor.
2. Portable wet/dry vacuum cleaner.
3. Sterile gauze sponges.
4. Sterile scalpel blades #22.
5. Surgical sutures: silk black braided.
6. Polyethylene tubing: I.D. size 0.86 mm, O.D. size 1.27 mm.
7. Clamps.
8. Surgical scissors: straight, sharp/sharp, blade length 32 mm.
9. Surgical forceps: straight, serrated.
10. Vein pick catheter introducer.
11. Sutures with reverse cutting: silk black braided, 135° curvature, 24 mm $^3/_8$ circle needle.
12. Seeker: stainless steel, bent-blunted end.

3 Methods

Follow IACUC guidelines for survival surgery on USDA-covered species. The surgery is a noninvasive, subcutaneous procedure.

3.1 Pre-operation

1. Cut surgical sutures 8 in. long. Three threads are needed per rabbit.
2. Cut catheters 11 in. long, fold one end, and clamp (one per rabbit; *see* **Note 3**).
3. Anesthetize rabbits with xylazine and ketamine or as directed by IACUC (*see* **Note 4**).
4. Shave the neck of the rabbit with an electric razor (*see* **Note 5**).
5. Treat the shaved neck with betadine antiseptic skin cleaner.
6. Place anesthetized rabbits on their backs with their heads towards the surgeon.

3.2 Operation

1. With a sterile scalpel blade, make a 2.5–4 cm incision on the skin of the neck, slightly to the left side of the trachea (*see* **Note 6** and Fig. 1a).
2. Cut the fascial/fat tissue with small sharp surgical scissors, slightly to the left of the trachea and avoiding the jugular vein, which will be on the left side of the incision (*see* **Note 7**).
3. Use scissors to cut away the membrane tissue under the fat, which covers the trachea and neck muscles (*see* **Note 8** and Fig. 1b).
4. With forceps, find the natural access between neck muscles to expose the left carotid artery and vagus nerve along the vertebrae (*see* **Note 9** and Fig. 1c).
5. Lift the carotid artery with forceps to separate it from the vagus nerve. Lift the carotid with the serrated side of one of the forceps (Fig. 1d). Do not pinch and pull, as this may result in damage to the carotid artery or the vagus nerve under it.
6. Place two forceps under the carotid artery (perpendicular to it) and pull in opposite directions along the length of the carotid artery to separate the membranes around it until 2–3 cm of the carotid artery is exposed (Fig. 1e).
7. Place two suture thread under the carotid, one at the top end (closest to the head) and the other at the bottom end (closest to the chest).
8. Tie *only* the top suture thread with a double knot (*see* **Note 10**).
9. Add a third (middle) suture thread close to the bottom thread and make a loose knot in it. Be careful to keep this loop loose (about 2 cm) and in between threads (*see* **Note 11**).

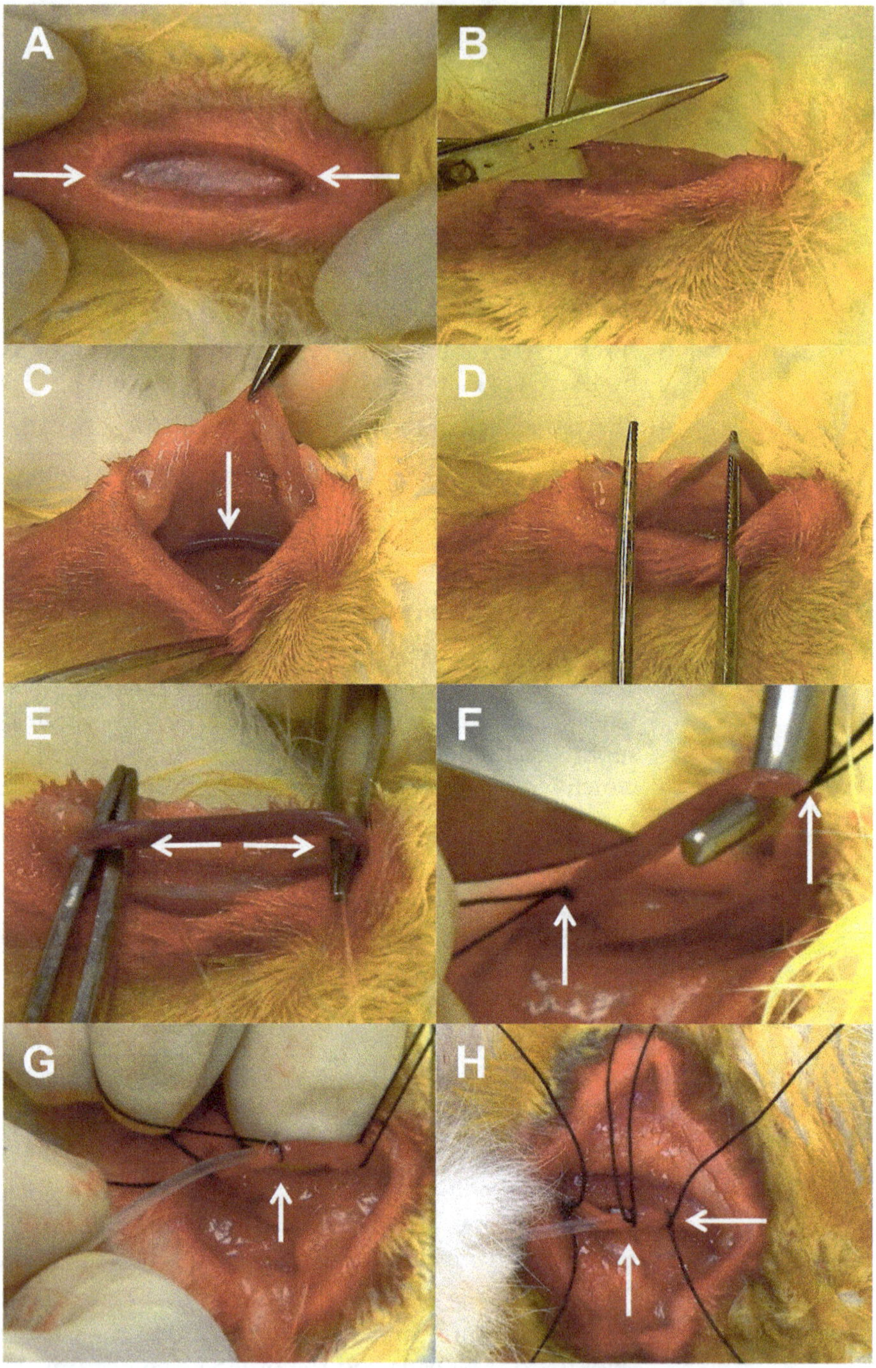

Fig. 1 Selected surgical procedure steps for catheterization of the left carotid artery in rabbits. *Panels* show area of the neck exposed for catheterization. The rabbit is positioned horizontally with the head on the *left side of the panel* and the torso on the *right side*. (**a**) An approximately 2.5 cm skin incision made adjacent to the trachea showing the underlying fascia and fat tissue. *Left arrow* points towards the top of the incision (closest to the head) and the *right arrow* points towards the bottom (closest to the chest). (**b**) Underlying tissue was lifted with forceps, a nick was made with small scissors, and the bottom scissor inserted while lifting to have

3.3 Catheterization

1. Requires two persons (*see* **Note 12**).
2. Assistant, lift the bottom suture thread (nearest the chest cavity) to stop blood flow to the carotid artery until the surgeon inserts the catheter in the exposed section of the artery.
3. Surgeon, with a sterile scalpel blade, nick the carotid artery parallel to the length of the carotid, not across as if cutting the artery (Fig. 1f). You will know the nick has been made when the blood remaining in the carotid leaks out. The assistant is critical here to prevent blood spurts from the artery (*see* **Note 13**).
4. Use a vein pick catheter introducer to find the nick and hold it for the introduction of the catheter.
5. With the free hand, insert the catheter into the hole (pointing towards the chest). Once the catheter is about 2–3 mm into the carotid artery, remove the vein pick catheter introducer and introduce the catheter until it reaches the threads held by the assistant to stop the blood flow. Carefully (to avoid the catheter from sliding back out of the carotid artery) tie off the looped thread snugly onto the catheter.
6. With one hand, hold the middle threads and with the other hold the catheter (Fig. 1g). When ready, instruct the assistant to drop the bottom sutures and gently push the catheter into the carotid about 8 cm to the heart. When the catheter reaches the heart, there will be mild resistance and then it will pulse when untouched. At that point the catheter is considered completely inserted and the bottom thread is tied off snugly onto the catheter (*see* **Note 14** and Fig. 1h).
7. Place a sterile gauze sponge soaked in sterile PBS over the incision site to protect it.
8. Let the catheter pulse against the heart valve for two additional hours. Record each animal catheter insertion time to be sure that each is in fact catheterized for 2 h (*see* **Note 15**).

Fig. 1 (continued) clear visibility of tissue being cut. (**c**) Left carotid artery is found in between the muscles under the membranous tissue cut in *panel* (**b**). (**d**) The left carotid artery is lifted with the serrated edge of one side of the forceps. (**e**) Two forceps were inserted under the carotid artery and pulled in opposite directions along the length of the carotid artery to separate the fascia around it. (**f**) The top end of the carotid artery (*left side*) was double-knotted, while two threads were placed on the bottom end (*right side*). One thread is being pulled by the assistant to stop blood flow to the exposed segment of the carotid, and the other was tied on a loose loop and left untouched at the moment. A nick is made on the carotid artery with a sterile blade. (**g**) Catheter was inserted into the carotid artery. *Arrow* points to the loose loop now tied snugly onto the catheter with the surgeon holding onto the threads. The assistant will drop the bottom threads and the surgeon will push the catheter in the direction of the bottom threads. (**h**) Left carotid artery after catheterization. Both the middle and bottom threads were tied off snugly onto the catheter. Pulsation of the catheter ensures correct placement

9. At the end of 2 h, remove the catheter (*see* **Note 16**). Gently pull the catheter out while holding the threads; it will come out fairly easily. When it pulls free of the bottom suture thread, stop and tie in a double knot but make sure that it is not so tight that it would break the carotid. Now the catheter can be completely removed and the middle thread double knotted (*see* **Note 17**).
10. Cut off excess suture threads, leaving about one-half inch on each end. The rabbit will recover fine with brain circulation coming from the right carotid.
11. Suture the incision site with reverse cutting sutures, placing a suture about every half centimeter.
12. Inject microbes at desired density, 2 ml per rabbit in PBS, in the marginal ear vein.

4 Notes

1. While having two persons makes it possible to perform the procedure, it is our advice that at least three persons be involved. The third person could be in charge of taking notes regarding time of administration of anesthesia and insertion of catheters and assessing the well-being of the rabbits while surgery is performed.
2. A 25-gauge, $^{7}/_{8}$ needle is best for the administration of anesthetics subcutaneously. Use a 26-gauge, $^{1}/_{2}$ needle if medication will be administered intramuscularly.
3. Cut catheters with a square edge. A clean cut will produce soft edges and will reduce the risk of damaging the artery as the catheter reaches the heart.
4. Calm rabbits (2–3 kg in size) with injection of xylazine. The same rabbits are then immediately injected with ketamine to anesthetize them.
5. Commercially available electric razors for pet grooming.
6. This is a blind cut. Hence, cutting next to the trachea ensures that the jugular vein will not be accidently cut.
7. Lift the fat tissue with forceps and cut straight through. The fat tissue is well vascularized in some rabbits, especially older ones. Hence, some self-contained bleeding may occur.
8. Lift the membrane underlying the fat, nick closer to the top of the incision (closer to the head), insert bottom scissor, and cut (Fig. 1b). The surgeon needs to always be aware of the tissue being cut to avoid unnecessary damage and excessive bleeding.
9. The carotid artery is about 1–2 mm in diameter and easy to see. The vagus nerve is bright white and must be separated

away from the carotid. Do not cut the vagus nerve or damage it or the animal will die.

10. Tying the top suture blocks circulation, allowing nicking and catheterization.
11. This third thread is to keep the catheter in place before the assistant lowers the bottom thread for insertion of the catheter into the heart and to avoid bleeding as the catheter is inserted. If bleeding is observed during insertion of the catheter, this thread should be tied a bit tighter.
12. Catheterization requires an assistant who will hold the bottom threads (nearest the chest cavity) to block blood flow, and the surgeon, who will nick the carotid artery and introduce the catheter.
13. The carotid will of course collapse and the nick will be harder to see at this point. The thread holder can lower the thread gently to help relocate the nick with the vein pick catheter introducer (cannula).
14. Make sure not to tie the threads tightly onto the catheter. Otherwise the catheter will be difficult to remove.
15. Check the catheters for proper placement every 15–20 min, as rabbits with high pulse can push the catheter back and the catheter stops pulsing or does not pulse as vigorously as when originally placed.
16. We have experimentally determined that 2 h of catheter insertion with 1.27 mm catheters leads to reproducible aortic valve damage, resulting in consistent vegetation formation—the characteristic lesion of endocarditis.
17. To remove the catheter, be sure it is not tightly stuck to the rabbit. PBS can be used to loosen it if it gets stuck.

Acknowledgment

This work was supported by the University of Iowa Carver College of Medicine Startup to WSP and PMS. We thank Katarina Kulhankova for the pictures.

References

1. Fowler VG, Durack DT (1994) Infective endocarditis. Curr Opin Cardiol 9:389–400
2. Durack DT, Lukes AS, Bright DK (1994) New criteria for diagnosis of infective endocarditis: utilization of specific echocardiographic findings. Duke Endocarditis Service. Am J Med 96:200–209
3. Cheung AL, Yeaman MR, Sullam PM, Witt MD, Bayer AS (1994) Role of the sar locus of *Staphylococcus aureus* in induction of endocarditis in rabbits. Infect Immun 62:1719–1725
4. van Wamel W, Xiong YQ, Bayer AS, Yeaman MR, Nast CC, Cheung AL (2002) Regulation of *Staphylococcus aureus* type 5 capsular

polysaccharides by agr and sarA in vitro and in an experimental endocarditis model. Microb Pathog 33:73–79

5. Xiong YQ, Bayer AS, Yeaman MR, Van Wamel W, Manna AC, Cheung AL (2004) Impacts of sarA and agr in *Staphylococcus aureus* strain Newman on fibronectin-binding protein A gene expression and fibronectin adherence capacity in vitro and in experimental infective endocarditis. Infect Immun 72:1832–1836
6. Xiong YQ, Fowler VG, Yeaman MR, Perdreau-Remington F, Kreiswirth BN, Bayer AS (2009) Phenotypic and genotypic characteristics of persistent methicillin-resistant *Staphylococcus aureus* bacteremia in vitro and in an experimental endocarditis model. J Infect Dis 199:201–208
7. Chuang ON, Schlievert PM, Wells CL, Manias DA, Tripp TJ, Dunny GM (2009) Multiple functional domains of *Enterococcus faecalis* aggregation substance Asc10 contribute to endocarditis virulence. Infect Immun 77:539–548
8. Frank KL, Guiton PS, Barnes AM, Manias DA, Chuang-Smith ON, Kohler PL et al (2013) AhrC and Eep are biofilm infection-associated virulence factors in *Enterococcus faecalis*. Infect Immun 81:1696–1708
9. Schlievert PM, Gahr PJ, Assimacopoulos AP, Dinges MM, Stoehr JA, Harmala JW et al (1998) Aggregation and binding substances enhance pathogenicity in rabbit models of *Enterococcus faecalis* endocarditis. Infect Immun 66:218–223
10. Fan J, Zhang Y, Chuang-Smith ON, Frank KL, Guenther BD, Kern M et al (2012) Ecto-5'-nucleotidase: a candidate virulence factor in *Streptococcus sanguinis* experimental endocarditis. PLoS One 7:e38059
11. Nallapareddy SR, Singh KV, Murray BE (2008) Contribution of the collagen adhesin Acm to pathogenesis of *Enterococcus faecium* in experimental endocarditis. Infect Immun 76:4120–4128
12. Nallapareddy SR, Singh KV, Sillanpaa J, Garsin DA, Hook M, Erlandsen SL et al (2006) Endocarditis and biofilm-associated pili of *Enterococcus faecalis*. J Clin Invest 116: 2799–2807
13. Hindes RG, Willey SH, Eliopoulos GM, Rice LB, Eliopoulos CT, Murray BE et al (1989) Treatment of experimental endocarditis caused by a beta-lactamase-producing strain of *Enterococcus faecalis* with high-level resistance to gentamicin. Antimicrob Agents Chemother 33:1019–1022
14. Schlievert PM, Dunny GM, Stoehr JA, Assimacopoulos AP (1997) Aggregation and binding substances enhance pathogenicity in a rabbit model of *Enterococcus faecalis* endocarditis. Adv Exp Med Biol 418:789–791

Chapter 7

Rabbit Model for Superantigen-Mediated Lethal Pulmonary Disease

Kristi L. Strandberg, Jessica H. Rotschafer, and Patrick M. Schlievert

Abstract

Staphylococcus aureus is a highly significant cause of serious human infections in the USA. Many of these illnesses are mediated by interactions between the host immune system and staphylococcal superantigens (SAgs). Several of these severe staphylococcal infections are initiated in the lungs, making this an important site to study. Here, we describe the rabbit model for investigating the role of staphylococcal SAgs in pulmonary-associated lethal infection and intoxication.

Key words *Staphylococcus aureus*, Superantigen, Rabbit model, Pulmonary infection, Pulmonary intoxication, Toxic shock syndrome

1 Introduction

Staphylococcus aureus is a significant human pathogen that causes many different illnesses including potentially fatal infections such as toxic shock syndrome (TSS) [1]. TSS is caused by staphylococcal superantigens (SAgs). SAgs are exotoxins that stimulate massive cytokine production by both T lymphocytes and antigen-presenting cells (APCs) [2, 3]. These cytokines include TNF-β, IL-2, IFN-γ, TNF-α, and IL-1β [4]. SAgs bind to the variable region on certain beta chains of the T cell receptor (Vβ-TCR) and either/both of the α- or/and β-chains of MHC II molecules present on macrophages [2, 5]. By binding to both the Vβ-TCR and MHC II molecule, the SAg is able to cross-link these two molecules regardless of the antigen peptide being presented by MHC II. This cross-linking action stimulates both the T cell and APC to release their respective cytokines. This massive cytokine release leads to the development of many of the clinical features of TSS.

In addition to causing TSS, SAgs are suspected to be involved in other serious and often fatal infections such as purpura fulminans and necrotizing pneumonia [6–8]. In order to better elucidate the role of SAgs in staphylococcal pulmonary infections, a

Amanda J. Brosnahan (ed.), *Superantigens: Methods and Protocols*, Methods in Molecular Biology, vol. 1396,
DOI 10.1007/978-1-4939-3344-0_7, © Springer Science+Business Media New York 2016

rabbit model was established. The model described here summarizes a safe and effective way to administer SAgs or live bacteria to rabbit lung tissues without the generation of aerosols. To better understand the link between the lethality of pulmonary staphylococcal infection and SAgs, rabbits were immunized against the SAg of interest prior to administering SAg or live bacteria. Immunization is NOT a required step in using this model, but is described in more detail here for those who wish to incorporate such a study into their own experimental design.

2 Materials

2.1 Immunization of Rabbits (Optional Step)

1. Dutch belted rabbits (1.5–2 kg) were used for all animal experiments. The care of all animals should be in accordance with guidelines established by the Institutional Animal Care and Use Committee.
2. Incomplete Freund's adjuvant.
3. Phosphate-buffered saline (PBS): 0.005 M $NaPO_4$, pH 7.2, 0.15 M NaCl.
4. Purified staphylococcal enterotoxin (SE) C (or other Sag of interest) can be purchased or produced according to the method outlined in Chapter 2 in this volume.
5. 5 mL luer lock syringes.
6. 18- and 25-gage needles.
7. Betadine solution.
8. Sterile gauze pads.
9. Microcentrifuge.
10. Microcentrifuge tubes.
11. Sterile toothpicks (sterilize by autoclaving).

2.2 Anti-SEC Serum Titer Enzyme-Linked Immunosorbent Assay (ELISA; Optional Step: Perform Only if Immunizing Rabbits Prior to Challenge)

1. Purified SEC that has been resuspended in sterile water (at a final concentration of 1 mg/mL).
2. High-binding polystyrene 96-well plates (such as Nunc Maxisorp).
3. 0.05 M Sodium bicarbonate-carbonate buffer pH 9.6: Dissolve 1.59 g of sodium carbonate (Na_2CO_3) and 2.93 g of sodium bicarbonate (NaH_2CO_3) in approximately 800 mL of water. Adjust the pH to 9.6 with HCl and NaOH and then bring the total volume of the solution up to 1000 mL. Store buffer at 4 °C for up to 6 months. Allow buffer to come to room temperature prior to use.
4. PBS-Tween-FCS: To prepare 1 L of PBS containing 0.05 % Tween20 and 1 % fetal calf serum, open a fresh 1 L bottle of

PBS and add 500 μL of Tween20 and 10 mL of fetal calf serum. Store buffer at 4 °C for up to 1 month. Allow buffer to come to room temperature prior to use.

5. Horseradish peroxidase-conjugated anti-rabbit antibody.
6. Substrate solution (1 % OPD solution): Note that this buffer is made in sequential steps, as indicated below. All solutions should be prepared immediately before use. Use at room temperature.
   (a) Prepare 100 mL of citric acid solution by dissolving 21 g of citric acid in 100 mL water.
   (b) Prepare 100 mL of phosphate solution by dissolving 35.6 g of disodium hydrogen phosphate in 100 mL water.
   (c) Prepare citrate phosphate buffer by combining 41 mL of citric acid solution with 51 mL of phosphate solution and adjust pH to 5 with HCl and NaOH.
   (d) Prepare 1 % OPD substrate solution by adding 0.3 g *o*-phenylenediamine and 60 μL of 30 % H_2O_2 to 30 mL of citrate phosphate buffer.
7. Stop solution (12.5 % sulfuric acid): Dilute 12.5 mL of concentrated sulfuric acid into 87.5 mL of water (*always dilute concentrated acids into water, never add water to the concentrated acid*). Store at room temperature.

2.3 Pulmonary Infection/ Intoxication Model

1. Dutch belted rabbits (1.5–2 kg) were used for all animal experiments. The care of all animals should be in accordance with guidelines established by the Institutional Animal Care and Use Committee.
2. The pulmonary infection experiments shown in this publication utilized the *S. aureus* USA400 isolate MW2 [SEC+, α-toxin+, and Panton-Valentine leucocidin (PVL)+] [7, 9]. Additional studies have been performed using the USA200 strains MNPA, MN1021, and MN128 [toxic shock syndrome toxin (TSST)-1+, α-toxin-, β-hemolysin-, γ-hemolysin-, and PVL-], and the USA400 strain c99-529 (SEB+, α-toxin+, and PVL+) [9]. Refer to original publication for additional information regarding experiments involving the SAgs TSST-1 and SEB [9]. All bacteria are grown in Todd Hewitt broth or dialyzed beef heart medium.
3. Scalpel.
4. Scissors.
5. PBS: 0.005 M $NaPO_4$, pH 7.2, 0.15 M NaCl.
6. 1 mL luer lock syringes.
7. 18- and 25-gage needles.
8. Electric beard trimmer.

9. Polyethylene tubing with 1.27 mm external diameter.
10. Sutures.
11. Vetbond Tissue Adhesive.
12. Betadine solution.
13. Sterile gauze pads.
14. Ketamine: 25 mg/kg.
15. Xylazine: 20 mg/kg.
16. Beuthanasia D: 1 mg/kg.

3 Methods

3.1 Immunization of Rabbits (Optional Step)

Immunization of rabbits is NOT required for this pulmonary model, but is included here as a useful tool for determining whether or not the SAg is contributing to the severity/lethality of the infection or intoxication.

1. To prepare SEC (or your SAg of interest) for all injections (initial immunization and booster immunizations) resuspend in sterile PBS at a concentration of 100 μg/mL.
2. Mix the resuspended SEC with an equal volume of incomplete Freund's adjuvant.
3. To immunize, each rabbit will receive a total volume of 1.0 mL (50 μg/mL SEC) injected subcutaneously over four different injection sites (0.25 mL per injection site) located on nape/scruff of the neck and back. The immunization schedule is as follows: each rabbit received the initial 1.0 mL injection, followed by 1.0 mL booster injections every 2 weeks until antibody titers were >10,000 as determined by ELISA (refer to Subheading 3.1 for protocol; *see* **Note 1** for a detailed explanation of titers).
4. For antibody titer assessment, collect approximately 0.5 mL of blood from the marginal ear vein of each rabbit (both immunized rabbits and unimmunized control rabbits) and place into a microcentrifuge tube.
5. Allow blood to clot at room temperature for 2 h.
6. Scrape the clot off the wall of the microcentrifuge tube by stirring around the edge of the tube using a sterile toothpick.
7. Once the clot has been separated from the tube walls, incubate the serum at 4 °C for 24–48 h to help shrink the clot, making the serum easier to remove from the tube (*see* **Note 2**).
8. Carefully remove serum from tube using a micropipette and transfer to new microcentrifuge tube. Store at 4 °C until ready to perform titer ELISA.

3.2 Anti-SEC Serum Titer ELISA (Optional Step: Perform Only if Immunizing Rabbits Prior to Challenge)

1. To determine anti-SEC antibody titers of immunized rabbits, coat the wells of a flat-bottom 96-well plate with SEC by adding 100 μL of 10 μg/mL (1 μg/well) purified SEC [10] and then incubating overnight in a 37 °C incubator to dry down the well contents.
2. Once dry, gently wash wells three times with 300 μL PBS-Tween-FCS, and then tap plate on stack of paper towels to remove access wash buffer.
3. Add 180 μL PBS-Tween-FCS to the first column of wells, and add 100 μL PBS-Tween-FCS to the rest of the wells.
4. Add 20 μL of undiluted rabbit serum to the wells in column 1 (wells will contain 200 μL).
5. Perform twofold serial dilution across the plate (*see* **Note 3**).
6. Incubate the plate for 2 h at room temperature.
7. Gently wash wells two times with 300 μL PBS-Tween-FCS, and then tap plate on stack of paper towels to remove access wash buffer.
8. Add 100 μL horseradish peroxidase-conjugated anti-rabbit antibodies to each well, and then incubate plate for 2 h at room temperature.
9. Gently wash wells two times with 300 μL PBS-Tween-FCS, and then tap plate on stack of paper towels to remove access wash buffer.
10. During the final wash step, prepare fresh substrate solution. Keep in dark bottle (*see* **Note 4**).
11. Add 100 μL of substrate solution to each well, and incubate in the dark for 30 min at room temperature to allow color to develop.
12. Stop the color development reaction by adding 50 μL stop solution.
13. Measure the absorbance at 490 nm wavelength.
14. If serum titers show greater signal by ELISA than control (unimmunized) serum when diluted 1:10,000 or greater, then rabbits are considered "hyperimmune" and are ready for use in pulmonary infection or intoxication experiments (*see* **Note 5**). If titers do not show greater signal by ELISA, then rabbits require additional immunization boosters. An example of an experiment done using immunized versus unimmunized rabbits is shown in Fig. 1.

3.3 Pulmonary Infection Model

One day prior to operation

1. One day prior to intrapulmonary inoculation with *S. aureus*, overnight bacterial growth cultures need to be started. Use a sterile loop to aseptically transfer colony (from plate) or loopful

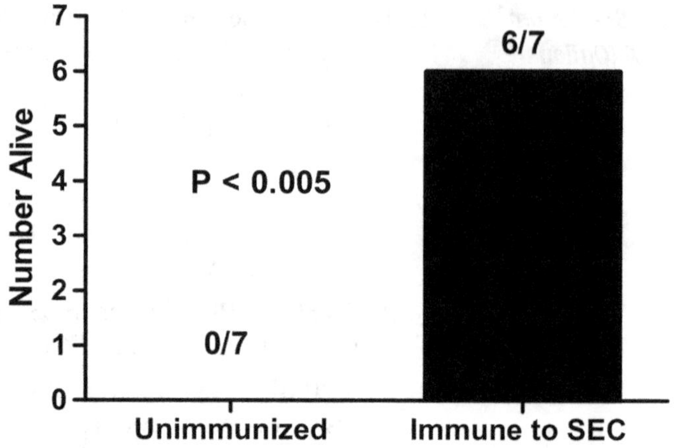

Fig. 1 Immunization against superantigen protects rabbits from developing fatal pulmonary infection. Number of non-immunized and SEC-immunized rabbits alive following challenge with live SEC+ USA400 (2×10^9 cells/mL) strain (reproduced from ref. 9 with permission from Oxford University Press)

(from broth culture) to a sterile 125 mL Erlenmeyer flask containing 25 mL of Todd Hewitt broth (or dialyzed beef heart medium) and then incubate overnight at 37 °C with shaking.

2. Rabbits will be inoculated with 2×10^9 cells in 200 μL culture medium. Prior to inoculation, adjust bacterial culture to appropriate cell density (*see* **Note 6**).

Pre-operation

1. Rabbits are anesthetized by intramuscular injections of ketamine and xylazine (*see* **Note 7**) [11].
2. Place each rabbit on its back, exposing its throat. Position the rabbit so that its head is pointing away from the surgeon and its feet are pointing towards the surgeon.
3. Using an electric beard trimmer, remove the fur from the rabbit's throat.
4. Purified SAg or live bacteria will be administered to rabbit lungs using polyethylene tubing. To prepare, cut a section of tubing that is long enough to be threaded down into the rabbit trachea while leaving several extra inches of tubing (approximately 8 in. long).
5. Fill a 1.0 mL syringe with either the purified SAg solution or the live bacterial culture.
6. Place an 18-gage needle on the end of the syringe and carefully insert the needle into one end of the tubing (*see* **Note 8**).
7. Once the syringe has been secured to the tubing, hold one end of the tubing up and then carefully depress the plunger on the syringe until liquid has filled the tubing line (but has not yet begun to spill out of the open end of the tubing).

8. Using a sterile gauze pad, apply betadine solution to the shaved area on the rabbit throat to disinfect the area.

Operation

1. Continue to keep the rabbit positioned on its back with its head pointing away from the surgeon. The trachea is very close to the surface when the rabbit is positioned in this manner. Using a finger, gently feel along the throat for the firm cartilaginous rings of the trachea (these will feel like horizontal rings that run perpendicular to the throat).
2. Once the trachea has been located, use a sterile scalpel to make a small incision (~1 in. long) just to the side (either the left or the right side is acceptable) of the trachea to cut through the skin.
3. Next, cut through the thin layer of muscle just beneath the skin. Take care not to cut the trachea itself.
4. Once the trachea has been exposed, use the scalpel to make a small hole (~3 mm) in the trachea. This hole can be made anywhere along the trachea as long as it is in between the firm cartilaginous rings. The exact location is not as important as being able to find the hole after it has been made so that the tubing can be inserted into that hole.
5. Gently insert polyethylene tubing through the 3 mm hole and thread down into the lung (*see* **Note 9**).
6. Administer bacteria or purified SAgs into the lungs through the tubing. Although the volume of bacteria or SAg being administered cannot be monitored directly as the end of the tubing is now inside the lung, it can be inferred by watching the liquid level change in the syringe as the plunger is depressed. Stop depressing the plunger once the desired volume has been administered.
7. Remove the tubing by gently pulling it out of the rabbit. Place the tubing somewhere safe to avoid accidental contact with any live bacteria or purified SAg remaining in the tubing.
8. Apply a drop of Vetbond Tissue Adhesive to the hole in the trachea to seal it. Wait for a minute or two to allow the Vetbond Tissue Adhesive to dry.
9. Suture the incision site, placing a suture about every 0.5 cm.

Post-operation

1. Carefully place the anesthetized rabbits in a safe area where they will not get injured as they awaken. The rabbits will be groggy and lethargic when they first wake up. Do not leave them on the surgical table, as there is a greater risk for them to be injured by falling off the table.
2. An analgesic can be administered by intramuscular injection once the rabbit is awake, if desired.

3. Wait until the rabbit is no longer lethargic before placing it back in its cage. There may be a greater likelihood of the rabbit injuring itself on something in its cage if it is still lethargic.
4. Monitor rabbits for the development of fatal respiratory distress and TSS (*see* **Note 10**). Animals that develop signs of fatal respiratory distress or TSS are euthanized by intravenous administration of Beuthanasia D. Rabbits that do not develop signs indicative of fatal respiratory distress and TSS are euthanized after 7 days following intrapulmonary challenge.

Postmortem tissue removal (optional step): Remove lung tissues from all rabbits for further analysis. Lung tissues can be photographed to document their overall appearance (*see* Fig. 2 for examples) or fixed and stained for further analysis.

1. To remove the lungs, first open the chest cavity of the deceased rabbit. Feel along the abdomen for the bottom of the rib cage and then using a scissors or scalpel, make a cut just below the rib cage and cut upward (towards the head) along the center of the ribcage (following along the sternum).
2. Once the ribcage is exposed, carefully cut open the ribcage starting at the bottom and cutting upwards towards the head. Cut close to the center of the ribcage (along to the sternum). Take care to avoid accidentally damaging the lung tissue.
3. Locate where the bronchial tubes branch off from the trachea. The heart will be blocking the area where the bronchial tubes branch from the trachea. Hold the heart out of the way with a forceps to gain better access to the area to remove the lungs.
4. Carefully cut through the bronchial tube and remove the entire lung (*see* **Note 11**). Once one lung has been removed, repeat the procedure to remove the second lung.
5. Capture images of the lung tissue, if desired.
6. After images are captured, sections of lung tissue can be removed for histological processing (for example, *see* Fig. 3). For the histological analysis shown in Fig. 3, rabbit lung tissue samples were fixed in 10 % formalin and then embedded in paraffin wax. Tissues were cut into sections 10 μm thick and then stained using hematoxylin and eosin following standard protocols.

Fig. 2 (continued) right lungs removed from a rabbit that received 2×10^9 cells MW2 in 200 μL dialyzed beef heart medium. (**c** and **d**) Whole left and right lungs removed from a rabbit hyperimmunized against purified SEC prior to administration of 2×10^9 cells MW2 in 200 μL dialyzed beef heart medium. (**e** and **f**) Whole left and right lungs removed from a rabbit after pulmonary exposure to 200 μg purified SEC in 200 μL sterile PBS. (**g** and **h**) Whole left and right lungs removed from a rabbit hyperimmunized against purified SEC prior to administration of 200 μg purified SEC in 200 μL sterile PBS. (**i** and **j**) Whole left and right lungs removed from rabbit exposed to 200 μL sterile PBS (reproduced from [9] with permission from Oxford University Press)

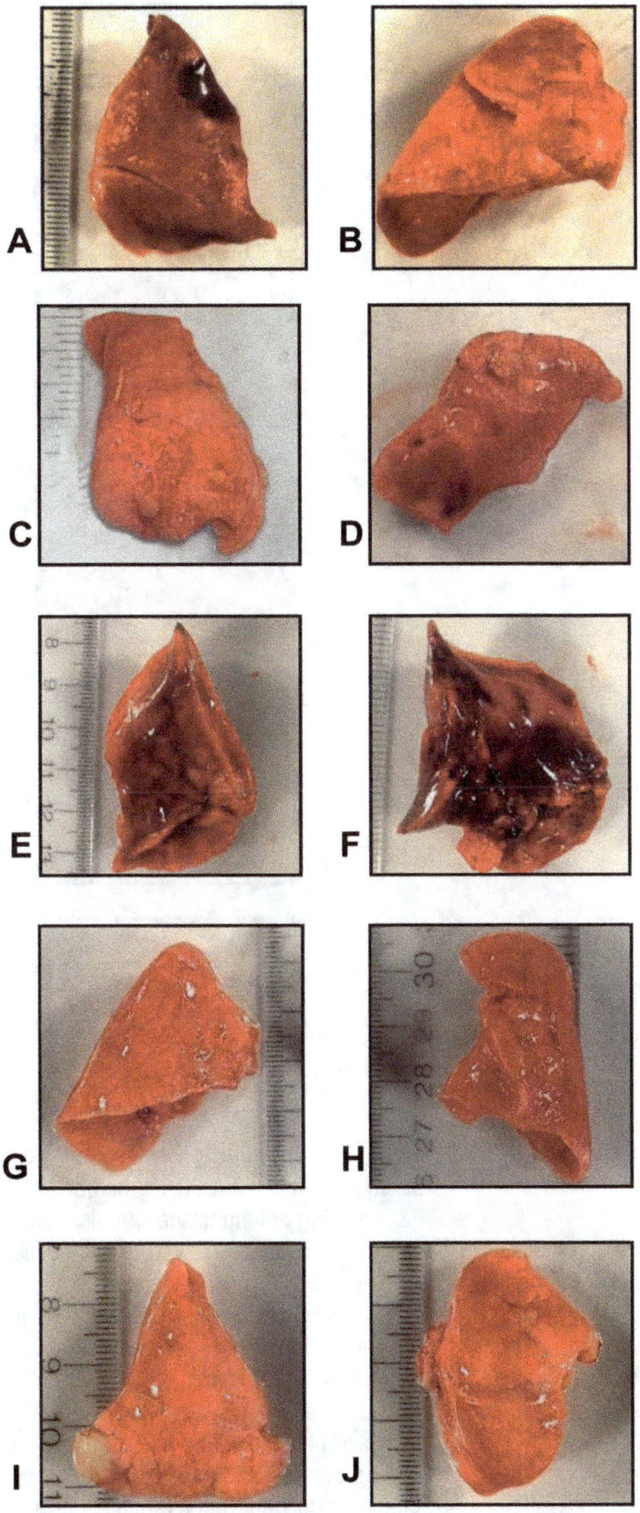

Fig. 2 Pulmonary disease of rabbit lung tissue after exposure to *S. aureus* strain MW2 (SEC$^+$, α-hemolysin$^+$, and PVL$^+$) or purified SEC. (**a** and **b**) Whole left and

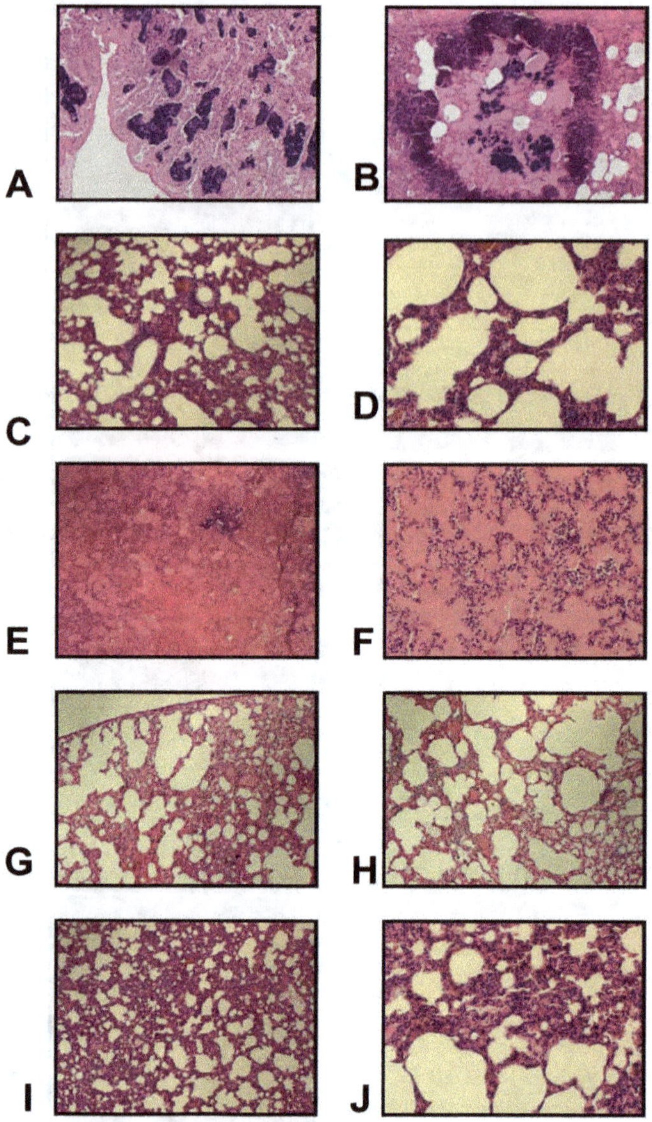

Fig. 3 Histology of lung tissue samples collected from rabbits challenged with *S. aureus* strain MW2 (SEC$^+$, α-hemolysin$^+$, and PVL$^+$), or purified SEC. (**a** and **b**) H&E staining of lung tissue samples removed from a rabbit (lung images shown in Fig. 2a, b) that received 2×10^9 cells MW2 in 200 μL dialyzed beef heart medium. (**c** and **d**) H&E staining of lung tissue samples removed from a rabbit hyperimmunized against purified SEC prior to administration of 2×10^9 cells MW2 in 200 μL dialyzed beef heart medium. (**e** and **f**) H&E staining of lung tissue samples removed from a rabbit after pulmonary exposure to 200 μg purified SEC in 200 μL sterile PBS. (**g** and **h**) H&E staining of lung tissue samples removed from a rabbit hyperimmunized against purified SEC prior to administration of 200 μg purified SEC in 200 μL sterile PBS. (**i** and **j**) H&E staining of lung tissue samples removed from a rabbit exposed to 200 μL sterile PBS (reproduced from [9] with permission from Oxford University Press)

4 Notes

1. The antibody titer is a measurement of how much antibody each rabbit has produced against purified SAg. Titer is expressed as the inverse of the greatest dilution that still gives a positive result. A common way to measure antibody titer is by ELISA. In this case the signal being measured is an absorbance value. When we state that an animal has a titer of 10,000, this means that the sample being tested still gave a positive result (greater signal) when diluted 1:10,000.
2. Serum is a component of blood. It is a yellow-colored, protein-rich clear liquid that separates out when blood coagulates. It does not contain white or red blood cells or clotting factors.
3. To perform twofold serial across the 96-well ELISA plate, use a multichannel pipette to gently mix the solutions in the wells of column 1 (pipette up and down 5–10 times to mix), and then transfer 100 μL from the wells in column 1 to the wells in column 2 (100 μL from well A1 is transferred to well B2, and so forth). Change tips in between each transfer. Once 100 μL has been transferred into the wells of column 12, use the multichannel pipette to carefully remove and discard 100 μL from all wells in column 12 so that the final volume in all wells across the plate is 100 μL.
4. The substrate solution is light sensitive and should be protected from direct sunlight and UV light. Storing this solution in a dark bottle (amber-colored glass, for example) will help protect it from sunlight exposure. Failure to protect this solution from light can result in decreased activity, which would result in an ELISA with less sensitivity.
5. To determine whether or not rabbits had been sufficiently immunized compare the absorbance values obtained from the unimmunized control serum with the absorbance values obtained from the immunized serum. If the absorbance values from the 1:10 diluted unimmunized control wells are less than the absorbance values obtained from the immunized serum (in the wells where the dilution factor is 1:10,000 or greater), then the anti-SEC titers are considered to be ≥1:10,000. At this point, the rabbits are considered "hyperimmune" and are ready for challenge with live bacteria or purified SAg.
6. For *S. aureus* strain MW2, an OD of 1.0 is equivalent to 1×10^9 CFU/mL; however, the cell density of any different strains should be determined experimentally prior to use in the pulmonary infection model.
7. Determine that the rabbit is under anesthesia by checking if the rabbit is able to right itself when placed on its back. A rabbit that is under anesthesia will not try to turn over and get back on its feet.

8. Depending on the interior diameter of the tubing used, you may need to use a different gage needle. The important thing is that the needle fits snuggly inside the tubing to minimize the risk of leaking. Always take care when working with needles to avoid sticking oneself. If an accidental needle stick occurs, seek medical attention immediately.

9. How to tell when the tubing is positioned in the lungs can be difficult. Feel for the bottom of the ribcage to help get a better idea of where the bottoms of the lungs are located in the rabbit. Keeping this in mind as the tubing is being inserted into the lungs will help the surgeon know how much tubing needs to be inserted to reach the lungs. As the tubing is being inserted, it will most likely encounter some resistance when it reaches the area where the trachea branches into the bronchial tubes. Although not required, it may be helpful to gently roll the rabbit onto its side (it does not matter which side) to help better guide the tubing down into that lung. Only one lung will receive the inoculation. Most likely it will be the lung located on the side that the rabbit was rolled onto, but this may not always be the case. As tubing continues to be inserted, a second (and less rigid) point of resistance will be encountered. This softer structure is lung tissue. Pull back slightly on the tubing to avoid further damage to the lung tissue and then administer the live bacteria or purified SAg. Once finished administering the challenge, pull the tubing out of the lung. Position the rabbit on its back to prepare for closing the incision (if the rabbit had been rolled onto its side).

10. The development of fatal respiratory distress or TSS corresponds to either death or, in agreement with the University of Minnesota IACUC and 28 years' research experience (PMS), failure of rabbits to exhibit both escape behavior and ability to right themselves. Rabbits typically show escape behavior by moving away from a person when he or she opens the cage door and tries to take the rabbit out of the cage. The ability to right themselves can be demonstrated by laying a rabbit down on its back or side. A normal or healthy rabbit will be able to get back on its feet (and will do so quickly). A moribund rabbit will not right itself quickly.

11. A considerable amount of blood may pool in the chest cavity once the first lung has been removed (depending on how recent the animal's passing was). If the chest cavity fills with blood, pause for a few minutes to allow the blood to clot, and then remove the clot by picking it up with a paper towel.

References

1. Lowy FD (1998) Staphylococcus aureus infections. N Engl J Med 339:520–532
2. Marrack P, Kappler J (1990) The staphylococcal enterotoxins and their relatives. Science 248:705–711
3. McCormick JK, Yarwood JM, Schlievert PM (2001) Toxic shock syndrome and bacterial superantigens: an update. Annu Rev Microbiol 55:77–104
4. Li H, Llera A, Malchiodi EL, Mariuzza RA (1999) The structural basis of T cell activation by superantigens. Annu Rev Immunol 17: 435–466
5. Wang L, Zhao Y, Li Z, Guo Y, Jones LL et al (2007) Crystal structure of a complete ternary complex of TCR, superantigen and peptide-MHC. Nat Struct Mol Biol 14:169–171
6. Daum RS, Ito T, Hiramatsu K, Hussain F, Mongkolrattanothai K et al (2002) A novel methicillin-resistance cassette in community-acquired methicillin-resistant *Staphylococcus aureus* isolates of diverse genetic backgrounds. J Infect Dis 186:1344–1347
7. Centers for Disease Control and Prevention (1999) Four pediatric deaths from community-acquired methicillin-resistant *Staphylococcus aureus*–Minnesota and North Dakota, 1997-1999. JAMA 282:1123–1125
8. Klevens RM, Morrison MA, Nadle J, Petit S, Gershman K et al (2007) Invasive methicillin-resistant *Staphylococcus aureus* infections in the United States. JAMA 298:1763–1771
9. Strandberg KL, Rotschafer JH, Vetter SM, Buonpane RA, Kranz DM, Schlievert PM (2010) Staphylococcal superantigens cause lethal pulmonary disease in rabbits. J Infect Dis 202:1690–1697
10. Hay FC, Westwood OMR (2002) Practical immunology. Blackwell Science, Malden, MA
11. Schlievert PM, Gahr PJ, Assimacopoulos AP, Dinges MM, Stoehr JA et al (1998) Aggregation and binding substances enhance pathogenicity in rabbit models of *Enterococcus faecalis* endocarditis. Infect Immun 66:218–223

Chapter 8

Nasopharyngeal Infection of Mice with *Streptococcus pyogenes* and In Vivo Detection of Superantigen Activity

Joseph J. Zeppa, Adrienne T. Wakabayashi, Katherine J. Kasper, Stacey X. Xu, S.M. Mansour Haeryfar, and John K. McCormick

Abstract

Streptococcus pyogenes is a globally prominent human-specific pathogen that is responsible for an enormous burden of infectious disease. Despite intensive experimental efforts to understand the molecular correlates that contribute to invasive infections, there has been less focus on *S. pyogenes* carriage and local infection of the nasopharynx. This chapter describes an acute nasopharyngeal infection model in mice that is utilized in our laboratory to study the role of superantigen toxins in the biology of *S. pyogenes*. We also describe a method to detect superantigen-specific T cell activation in vivo.

Key words *Streptococcus pyogenes*, Group A *Streptococcus*, Nasopharyngeal infection, Transgenic mouse model, Superantigen, Vβ skewing

1 Introduction

Streptococcus pyogenes [(commonly referred to as the group A *Streptococcus* (GAS)] is a globally prominent, human-specific pathogen. This organism contributes to over half a million deaths each year worldwide that are attributed primarily to invasive infections and post-infection autoimmune disorders such as rheumatic heart disease [1]. However, the most common forms of GAS-host interaction are relatively mild infections including pharyngitis, and asymptomatic carriage in the nasopharynx and on the skin. Indeed, there are over 600 million cases of pharyngitis worldwide each year, and up to 14 % of some populations are colonized asymptomatically by GAS [1, 2]. Despite the high frequency of the carriage state, and its probable importance for driving GAS transmission [3, 4], most experimental infection models have focused on severe and invasive diseases [5].

We recently described a *S. pyogenes*nasopharyngeal infection model in transgenic mice that express human leukocyte antigen

Amanda J. Brosnahan (ed.), *Superantigens: Methods and Protocols*, Methods in Molecular Biology, vol. 1396,
DOI 10.1007/978-1-4939-3344-0_8, © Springer Science+Business Media New York 2016

[HLA]-DR4 and HLA-DQ8 in a C57BL/6 background [6]. This model was adapted from an earlier infection model that located GAS colonization in the nasal associated lymphoid tissue (NALT) of BALB/c mice [7]. Our work demonstrated that efficient infection of the nasopharynx by the rheumatic fever isolate *S. pyogenes* MGAS8232 [6, 8] required both the expression of the streptococcal pyrogenic exotoxin A (SpeA) superantigen and the expression of HLA class II antigens. Herein, we describe our current infection procedure, including the procurement of a tissue complex we refer to as the complete nasal turbinates (cNT) containing the NALT, nasal turbinates, and maxillary sinuses [6]. Additionally, as there is still only limited evidence that demonstrates expression of these toxins during in vivo infection by GAS, we describe a flow cytometry-based method that evaluates TCR Vβ-specific activation as a measure of SpeA activity in vivo. Although this method focused on detection of SpeA activity, and thus examines the TCR Vβ8-expressing T cells targeted by SpeA [9], it could potentially be applied to assess the in vivo expression of other superantigens by examining the appropriate Vβ-specific T cell subpopulations.

2 Materials

Prepare all solutions using ultrapure water (deionized water sensitized to 18 MΩ cm at 25 °C). Store all reagents at room temperature unless otherwise stated. Follow hazardous waste disposal regulations according to your institutional policies.

2.1 Murine Infection

1. Prepare *Streptococcus pyogenes* according to Kasper et al. [6] (*see* **Note 1**).
2. Human leukocyte antigen (HLA)-DR4/DQ8-expressing mice [10] or other mouse strains deemed appropriate (*see* **Note 2**).
3. Hanks' balanced salt solution (HBSS).
4. Anesthesia apparatus: Anesthesia machine (Benson Medical Industries Inc.), 2000 L oxygen tank with a flow rate gauge, murine capacity gas anesthesia chamber with appropriate filter, and filtered nose cone.
5. Forane (Isoflurane, USP) inhalation anesthetic (Baxter Corporation).
6. Bench coat cut to 25 cm by 15 cm.

2.2 Lymph Node and cNT Extraction

1. 70 % Ethanol in water.
2. Sterile gauze.
3. Anesthesia and apparatus as above.

4. Surgical platform: Styrofoam platform (25 cm by 15 cm) that is sterilized using a chloride dioxide-based sterilant, lined with bench coat cut to similar size.
5. Surgical equipment: "T" pins (10×), 4.5 inch iris scissors with straight blades, 4.5 inch iris scissors with curved blades, 5.5 inch operating scissors with straight blade, 4.5 inch stainless steel curved splinter thumb serrated forceps, and 4.5 inch stainless steel curved splinter thumb forceps without serration.
6. Needles (25 gauge).
7. Heparin sodium 1000 USP units per mL.
8. Phosphate-buffered saline (PBS): 137 mM sodium chloride, 2.7 mM potassium. chloride, 10 mM sodium phosphate dibasic, 2 mM potassium phosphate monobasic; pH adjusted to 7.5 with concentrated sodium hydroxide.
9. 1.5 mL Eppendorf tubes containing 500 μL HBSS.

2.3 Complete Nasal Turbinate Processing

1. Glass homogenizer (1 mL).
2. HBSS.
3. Trypticase soy agar with 5 % sheep blood.
4. 37 °C static incubator.

2.4 Flow Cytometry Analysis

1. Glass homogenizer (5 mL).
2. PBS.
3. Cell strainer (40 μm Nylon).
4. FACS wash buffer: PBS + 2 % fetal bovine serum (heat inactivated), optional 0.01 % sodium azide for long-term storage (>2 months) stored at 4 °C after preparation.
5. Fc block obtained from eBioScience: Hybridoma clone 2.4G2 supernatant containing anti-CD16/CD32 monoclonal antibodies and stored at 4 °C.
6. Antibodies were obtained from eBioScience: α-CD3-Allophycocyanin (APC) (clone 145-2C11), α-CD25-phycoerythrin (PE) (clone 7D4), α-Vβ8.1/Vβ8.2 TCR-fluorescein isothiocyanate (FITC) (clone KJ16), α-Vβ3 TCR-FITC (clone KJ25) stored in the dark at 4 °C.

3 Methods

3.1 Murine Nasopharyngeal Infection

All work must be done in a Certified Animal Research Facility under Biosafety/Containment Level 2 standards by trained personnel. Mouse infections are performed in a Certified Biological Safety cabinet. All work must be approved by the relevant Institutional Animal Care and Use Committee. Mice are housed in microisolator cages; cage waste is treated as a biohazard and is incinerated.

1. Live *Streptococcus pyogenes* cells are prepared to a final concentration of 1×10^8 colony-forming units (CFU) per 15 μL of HBSS (*see* **Note 1**). This preparation will be inoculated into the nostrils of the HLA-DR4/DQ8 transgenic mice.
2. An individual mouse is anesthetized in an anesthesia chamber using a 4 % isoflurane mixer with oxygen at a flow rate of 1 L per minute. Mice are considered fully anesthetized when they no longer respond to the toe pinch maneuver (*see* **Note 3**).
3. While the mouse is undergoing anesthesia, prepare for the infection by drawing up 7.5 μL of the bacterial inoculum using a P10 or P20 pipetter. Of critical importance to the model, as shown by others [11], and our own unpublished observations, administration of larger volumes will result in bacterial pneumonia and thus does not represent a nasopharyngeal infection.
4. Once anesthetized, the mouse is scruffed and, as quickly as possible, the operator should apply the bacterial solution to one nostril of the mouse dropwise, allowing each drop to be inhaled before the next is applied (Fig. 1). Do not touch the

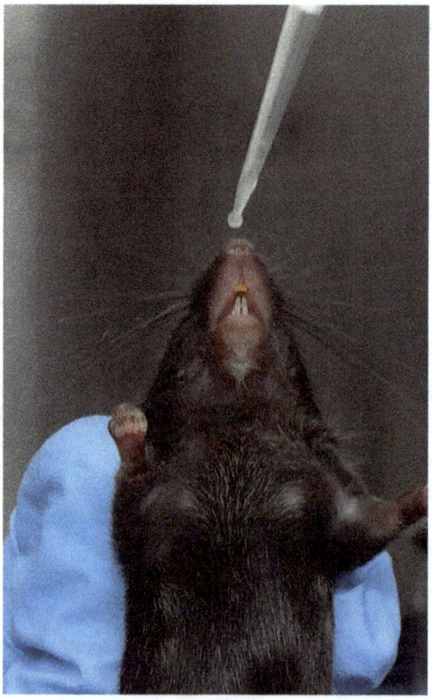

Fig. 1 Technique for nasopharyngeal infection of mice with *S. pyogenes*. Mice are anesthetized with a mixture of 4 % isoflurane with oxygen, at a flow rate of 1 L/min. Once fully anesthetized, mice are scruffed and inoculated with 7.5 μL per nostril with 1×10^8 CFU/15 μL of bacterial suspension, with anesthesia re-administered between each inoculation

nose of the mouse with the pipette tip. Immediately after the entire inoculum has been consumed, the mouse should be returned to the isoflurane chamber, belly up, to allow for the mouse to be re-anesthetized and prevent the inoculum from dripping out.

5. Prepare the second dose of the inoculum as in **step 3**, and once the mouse is again completely anesthetized [as determined by the toe pinch maneuver (*see* **Note 3**)], repeat the inoculation procedure to the remaining nostril as in **step 4**. Return the mouse to the anesthesia chamber until it is again fully anesthetized.
6. Move the animal to a recovery area (the lid of the animal cage with a piece of bench coat; this is to prevent inhalation of bedding material during recovery). Once recovered, the mouse can be moved back into its cage.
7. If only bacterial quantification from the nasopharynx is required, conduct Subheadings 3.2, **steps 1–4**, to exsanguinate the animal and then proceed to Subheading 3.4.

3.2 Cervical Lymph Node Extraction

Sterilize all surgical instruments by autoclaving prior to starting the procedure. Between each step, instruments should be sterilized with 70 % ethanol and wiped down with sterile gauze to minimize contamination.

1. Once the mice have been infected for 48 h (or other times as determined appropriate), fully anesthetize the mice (as in Subheading 3.1, **step 2**). Set up the anesthesia machine with chamber (as previously noted) as well as with a nose cone also connected to the isoflurane machine and a sterile surgical platform plus bench coat ready to receive the mouse.
2. Once the mouse is anesthetized in the chamber, divert the flow of isoflurane/oxygen from the chamber to the nose cone and move the mouse onto the surgical platform so that the mouse is in a ventral position and has its nostrils in the nose cone. Ensure that the mouse is fully anesthetized as determined by the toe pinch maneuver (*see* **Note 3**). Without puncturing the limbs, use "T" pins to secure both the arms and legs of the mouse to the platform in an "X" orientation (Fig. 2a). Sanitize the abdomen using 70 % ethanol.
3. Using serrated forceps, tent the skin of the abdomen and make an incision into the abdominal cavity (through the fur, dermis, and peritoneum) using straight blade iris scissors, being careful not to disrupt any major tissues or organs. Continue with the incision upwards until you reach the rib cage, and downwards towards the genitalia. Lateral incisions may also be made from the midpoint of the abdomen to allow for better access to the abdominal cavity.

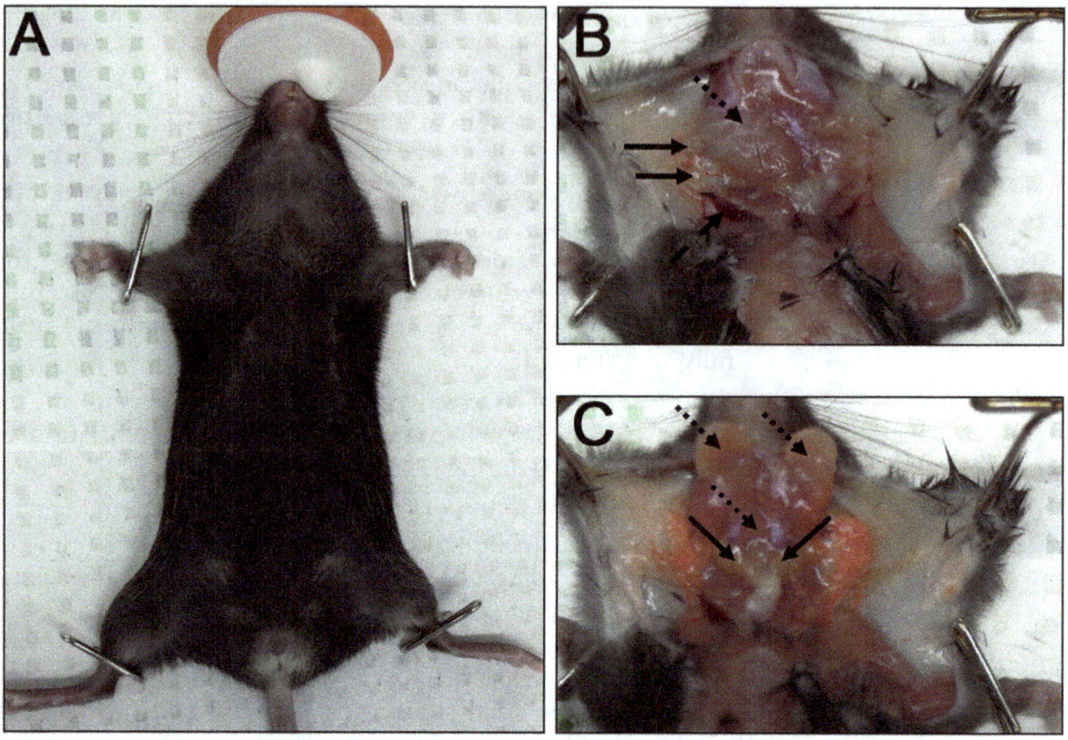

Fig. 2 Location of cervical and deep cervical lymph nodes. The fully anesthetized mouse should be positioned belly up with nose inserted into the nose cone. (**a**) The nose cone should be emitting a 4 % isoflurane mixture with oxygen at a 1 L/min flow rate. Legs and arms should be secured with "T" pins in an "X" orientation and the mouse confirmed to be anesthetized by the toe pinch maneuver (*see* **Note 2**). (**b**) Postmortem, incisions are made from below the rib cage to the chin. Diagonal incisions are also made up to the paws. The skin is peeled back to reveal the lymph nodes (*solid arrow*), external jugular vein (*rectangle-dashed arrow*), and the submaxillary gland (*square-dashed arrow*). (**c**) Deep cervical lymph nodes (*solid arrow*) are located beneath the submaxillary gland (*square-dashed arrow*) and next to the trachea (*circular-dashed arrow*)

4. In the opened abdominal cavity, move the intestines and any fat pads to the side to expose the inferior vena cava, being sure not to damage any of the internal organs or tissues in the process. Insert a heparin-primed 25-gauge needle as parallel as possible into the inferior vena cava and slowly draw back on the plunger until as much blood is procured as possible (*see* **Note 4**). Dispose of the syringe and needle in a sharps waste container. Sever the diaphragm located above the liver and wait until breathing and heartbeats stop. The isoflurane and oxygen can now be shut off.
5. Continue the upper incision that was left off at the rib cage making sure that only the skin is being cut. Continue until you reach the jaw. Also, create branch incisions that continue up diagonally towards the paw on each side of the mouse.
6. Ensure that the mouse is securely pinned to the surgical platform. Using serrated forceps peel the newly generated skin flaps

between each paw and the neck and secure them to the surgical platform using "T" pins. The trachea, submaxillary gland, sublingual gland, and cervical lymph nodes should now be exposed (Fig. 2b). Using curved, non-serrated forceps, collect the cervical lymph nodes into a sterile tube containing 5 mL of PBS and kept on ice until processing (*see* **Note 5**).

7. Deep cervical lymph nodes are obtained by lifting and removing the submaxillary gland (Fig. 2c). The lymph nodes are located underneath this tissue (*see* **Note 5**). Place the deep cervical lymph nodes into a sterile tube containing 5 mL PBS on ice until processing.

3.3 Cervical Lymph Node Processing for Analysis of Vβ-Specific T Cell Activation

To detect superantigens in vivo, a control infection is recommended with a strain of *S. pyogenes* lacking superantigens that target the Vβ subset of interest. Complete the following steps that involve antibody-fluorophore conjugates in as low light as possible.

1. Decant lymph nodes in PBS into a 5 mL glass homogenizer and homogenize. Decant tissue into a 15 mL conical tube and spin for 5 min at $330 \times g$.
2. Remove supernatant by decanting so that there is a small volume remaining in the tube and resuspended by racking the pellet (*see* **Note 6**). Add 5 mL of PBS and pour the suspension through a cell strainer to remove any debris. Wash the conical tube with 5 mL PBS, pour through the cell strainer, collect all strained cells in PBS, and spin as indicated above.
3. Remove supernatant and resuspend cells in 3 mL PBS. Enumerate live cells using a hemocytometer with trypan blue dye exclusion.
4. For each flow cytometry sample (unstained, single, double, and triple stained) aliquot 3.5×10^5 cells into a FACS tube. Wash cells by adding 2 mL of FACS buffer to each tube and spin at $330 \times g$ for 5 min. Remove supernatant, rack cells, and wash cells twice more for a total of three washes.
5. Remove supernatant from the last wash, rack cells, and add 40 μL of FC block to each tube. Leave on ice or at 4 °C for 20 min.
6. Add the manufacturer's recommended quantity of the indicated antibodies (a wash step between FC block and antibody addition is not required):

   The following antibody-fluorophore conjugates were used: CD3-APC, CD25-PE, Vβ8.1/Vβ8.2-FITC, and Vβ3-FITC. Unstained, single-stained, double-stained, and triple-stained samples are required for spectral compensation and analysis. The samples that will be analyzed for the experimental outcome will be the triple-stained $CD3^+CD25^+V\beta8.1/V\beta8.2^+$ cells and the $CD3^+CD25^+ V\beta3^+$ cells.
7. Once the appropriate antibodies have been added to the samples, incubate at 4 °C for 30 min in the dark.

8. Wash as in **step 4** for a total of three washes.
9. Resuspend cells in 400 μL of FACS wash buffer. The samples are now ready to be read on an appropriate flow cytometer.
10. The flow cytometer should be run according to the operator's manual. Ensure that all compensation is completed correctly according to the cytometer's software using the unstained samples, single-stained samples, and fluorescence-minus-one (FMO) controls.
11. To determine T cell Vβ-specific alterations, first gate on the live lymphocyte population based on forward and side scatter (Fig. 3a).
12. Next, gate on the CD3$^+$ cell population (Fig. 3b).
13. Lastly, analyze the CD3$^+$ population by gating on the Vβ8.1/Vβ8.2$^+$-FITC channel and the CD25$^+$-PE channel (Fig. 3c). As CD25 (i.e., the α-chain of the interleukin-2 receptor) is expressed on activated T cells, increases in cells that are double positive, compared with control conditions, are indicative of the activated Vβ subset of T cells targeted by the superantigen of interest.
14. This gating strategy is repeated with the Vβ3$^+$-FITC population in both wild-type and SAg knockout infections to observe any overall changes to the T cell population, thus acting as an internal T cell control (Fig. 3c).
15. Comparison is completed to determine the percentage of the activated population (CD25$^+$) compared to the entire CD3$^+$ Vβ$^+$ population of the superantigen-responsive subset (Vβ8 in the case of SpeA). Now examine the control Vβ population (vβ3) in the same manner, which indicates whether the infection in general is affecting the T cell population. Ideally, these should then be compared between the superantigen-expressing and superantigen knockout strains of *S. pyogenes*. Depending on the number of strains being compared, choose the appropriate statistical method. Figure 3d shows the comparison of wild-type *S. pyogenes* MGAS8232 with *S. pyogenes* MGAS8232 ΔSpeA analysis for both the total Vβ8-specific T cells that are targeted by SpeA, and Vβ3 as an internal control population, and Fig. 3e shows this comparison with the inclusion of the CD25 activation marker. This analysis demonstrates that although there are no significant differences in the percentage of Vβ3 or Vβ8-specific T cells, SpeA production by wild-type *S. pyogenes* induces Vβ8-specific T cell activation in vivo.

3.4 Complete Nasal Turbinate Processing and Bacterial Enumeration

1. Using serrated forceps, grip the cheek of the mouse and decapitate the mouse using the straight blade operating scissors (*see* **Note 7**) (Fig. 4a). The body of the mouse can now be disposed of.

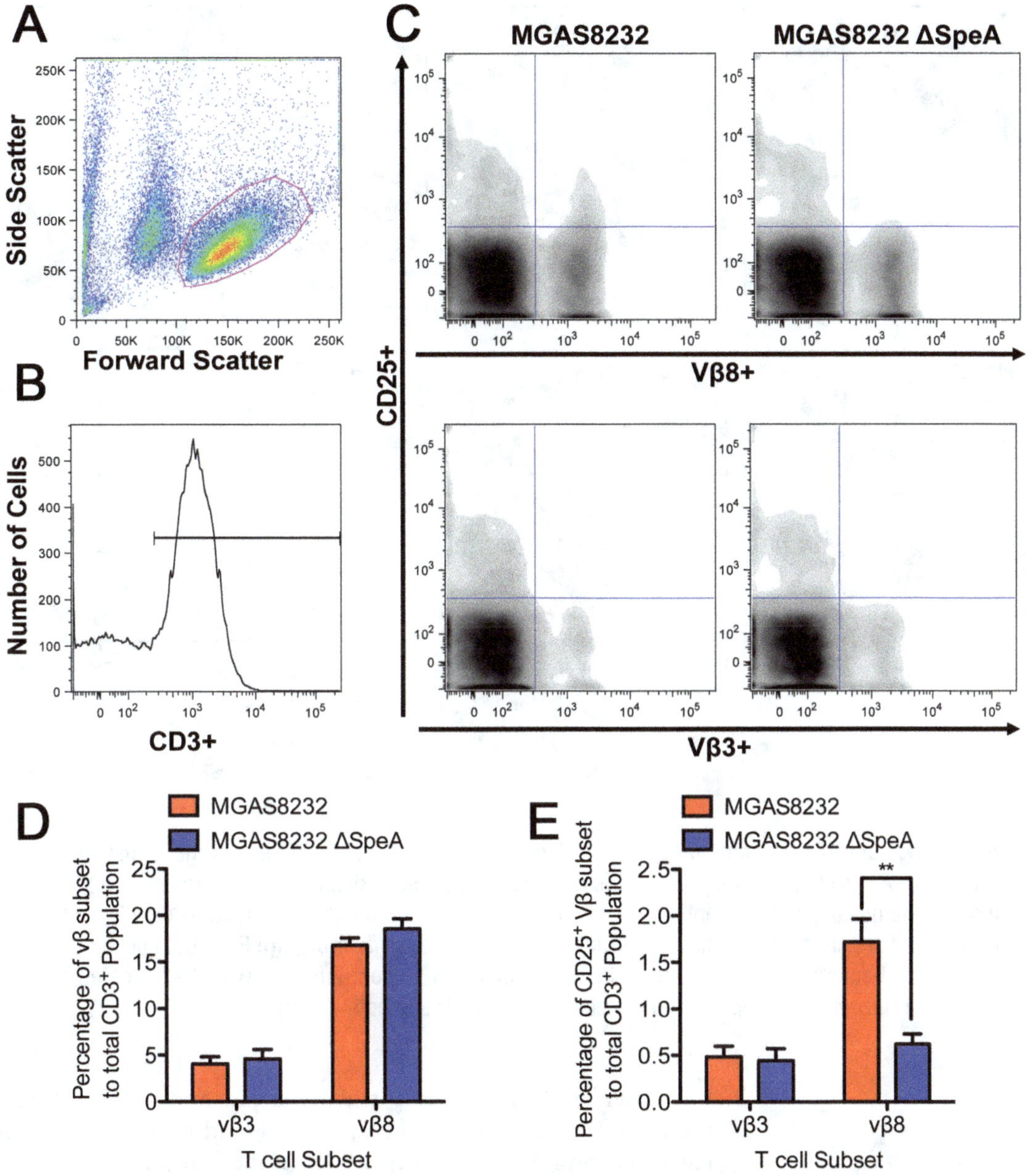

Fig. 3 In vivo detection of superantigen-dependent Vβ-specific T cell activation. (**a**–**c**) Flow cytometry analysis of HLA-DR4/DQ8 lymph node cells 48 h post-infection with *S. pyogenes* MGAS8232 wild type or MGAS8232 Δ*speA*. Representative flow cytometry gating strategy for (**a**) live lymphocytes, (**b**) $CD3^+$ cells (T cells), and either (**c**) $CD25^+V\beta8^+$ or $CD25^+V\beta3^+$ T cells. Staining for Vβ3 and Vβ8 was from the same mouse, with Vβ3 acting as the internal control for each mouse. Percentage of Vβ3- or Vβ8-specific subsets to total CD3+ T cells (**d**) or percentage of $CD25^+V\beta3$- or Vβ8-specific subsets to total CD3+ T cells (**e**) in wild-type *S. pyogenes* MGAS8232 ($n=4$) or MGAS8232 Δ*speA* ($n=5$) infected mice. Data are shown as mean ± standard errors of the means (SEM). Statistical significance is displayed as **, $p<0.01$ by unpaired, two-tailed Student's *t*-test

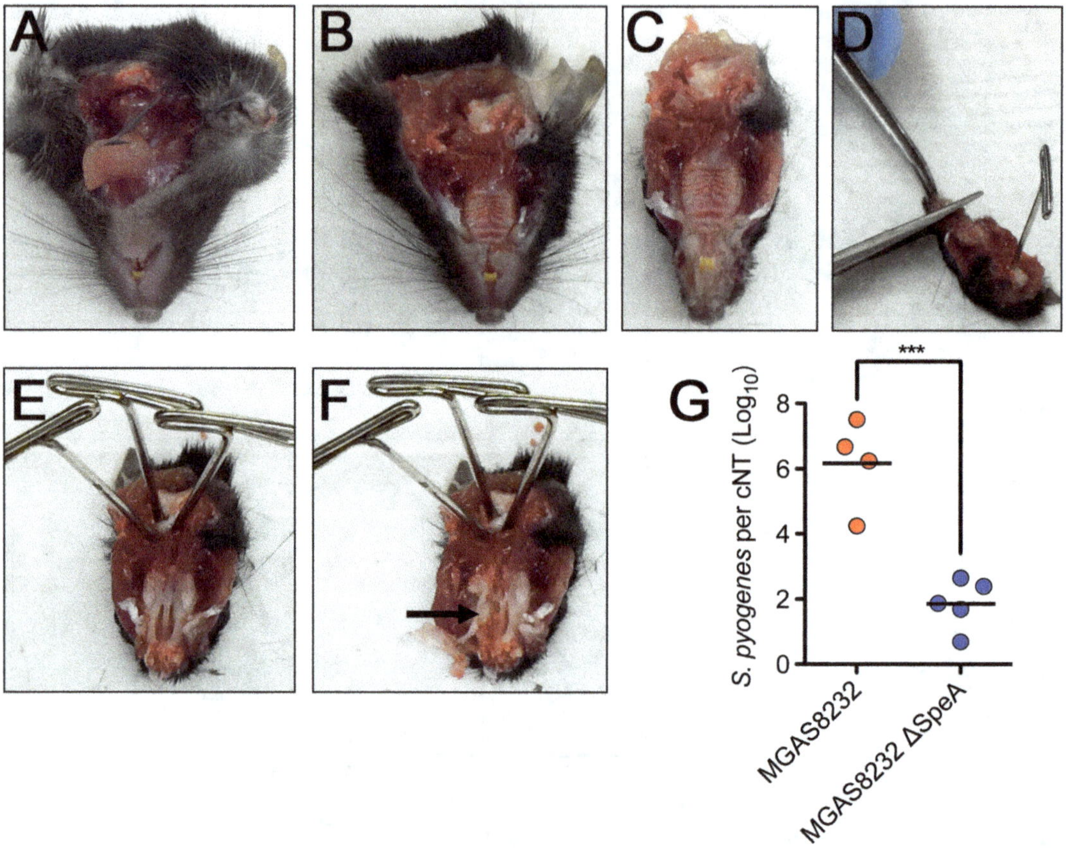

Fig 4 Extraction of cNT and enumeration of live *S. pyogenes*. The decapitated head of the mouse (**a**) has the lower jaw (**b**), excess cheek tissue (**c**), tip of the nose from the incisors (**d**) and the palate removed (**e**), exposing the maxilla, the molars, and the septum. (**f**) The complete nasal turbinates are located bilaterally posterior to the maxilla and are characterized as an off-white, mucoid tissue (*solid arrow*). (**g**) Each data point represents beta-hemolytic bacteria in the cNT from an individual mouse 48 h post-infection with the indicated strain of *S. pyogenes*. Horizontal bars represent the mean (***, $p < 0.01$; students' t-test)

2. Using serrated forceps, grip the lower jaw and tongue. Insert the operator scissors into the oral cavity and remove the lower jaw by cutting both connective sides of the mandibles (Fig. 4b). The lower jaw can now be disposed of.
3. Using the operator scissors, remove the excess tissue from both cheeks by gripping the loose skin with the serrated forceps and trimming. Do not remove any bone or cartilage during this process (Fig. 4c).
4. Flip the head so the molars and incisors are facing straight up and remove the tip of the nose by securing the head with a "T" pin between the visible external auditory meatus, gripping the nose with the serrated forceps and cutting just behind the incisors with the operating scissors (Fig. 4d).

5. Orient the head so that the area where the nose was is pointed towards the operator. Further secure the head with two additional "T" pins in the same area as before. Using serrated forceps, remove the hard and soft palate by simply peeling these tissues up and away from the skull (Fig. 4e). Dispose of the palate.
6. To gain access to the cNT, place one blade of an opened pair of curved iris scissors into the cavity between the left and right maxilla, and slide them up until the blades are beside the molars. Squeeze the scissors together without actually cutting into the upper jaw. Roll the scissors to the outside, breaking the molars away from their original position exposing the cNT (Fig. 4f).
7. The cNT is on the posterior side of the palate, bilaterally (Fig. 4f). Using a pair of smooth curved forceps, collect all of the off-white, mucoid tissue in the now exposed cavity into 500 μL of HBSS in a 1.5 mL Eppendorf tube. Repeat this procedure on the other side to procure the entire cNT (*see* **Note 8**). Store the cNT on ice until processing. Decant the cNT and HBSS into a sterile, 1 mL glass homogenizer. Thoroughly homogenize the tissue and decant the solution back into the container.
8. Serially dilute the homogenate with tenfold dilutions (using HBSS as the diluent) until a final dilution of 10^{-7} is reached. Plate two 10 μL drops of neat and each dilution on trypticase soy agar plus 5 % sheep blood (*see* **Note 9**). Also plate 100 μL of the neat homogenate on an individual plate. Plates are incubated at 37 °C for 48 h and then counted for bacterial enumeration. Figure 4g shows the bacterial quantitation for the Vβ-specific T cell activation data shown in Fig. 3.

4 Notes

1. *Streptococcus pyogenes* cultures are prepared as described in Kasper et al. [6]. Prior to murine infection, *S. pyogenes* liquid cultures are inoculated from −80 °C stock cultures [stored in Todd Hewitt medium supplemented with 1 % (w/v) yeast extract (THY) broth supplemented with 25 % glycerol] into sterile THY broth and subcultured (1 %) for 3 days before preparing the final inoculum of 1×10^8 CFUs in a total volume of 15 μL of HBSS.
2. We utilize transgenic mice expressing both HLA-DR4 and HLA-DQ8 on the C57BL/6 background as these mice are highly sensitive to the SpeA and SmeZ superantigens compared with wild-type C57BL/6 mice [6]. However, other investigators have utilized successfully FVB/n mice that are reported to

be more efficiently infected compared with C57BL/6, BALB/c, A/J, and CD1 mice [11].

3. The toe pinch maneuver ensures that the mouse is anesthetized entirely by testing its peripheral reflexes. Once the mouse appears to be anesthetized (i.e., is immobile), the operator can lightly pinch the digits of either of the mouse's feet. If the mouse responds with any involuntary contraction, wait for an additional 30 s before repeating. Once the mouse no longer responds to this stimulus, the nasal inoculation may proceed. If not entirely anesthetized, the mouse may sneeze out the bacterial inoculum and compromise the results of the experiment.
4. This procedure will exsanguinate the animal. Alternative approved exsanguinating procedures may also be used (such as the cardiac puncture by a trained individual). Exsanguination, as a euthanasia method, reduces the circulating blood volume resulting in cleaner decapitation and easier lymph node and cNT extraction.
5. This area of the mouse may sometimes contain large quantities of adipose tissue. To distinguish lymph nodes from fat, simply roll them along the bench coat on the surgical platform. If they hold their shape, they are lymph nodes; if not, they are adipose tissue.
6. Racking is achieved by running the pelleted tube against the bars of a tube rack to gently break up the pellet. We have not observed negative effects on the pelleted cells using this procedure.
7. After decapitation, some movements may be observed (i.e., gasping-like behavior). This is normal, involuntary, and not indicative of improper sacrifice.
8. At times, during the "roll" maneuver, the cNT sticks to the underside of the maxilla, so this area should be inspected for cNT. There can also be cNT attached to the septum, so this tissue should also be inspected.
9. Trypticase soy agar plus 5 % sheep's blood is a differential medium due to the presence of Streptolysin S production by *S. pyogenes*, which will generate a zone of clearance around GAS colonies. Todd Hewitt containing 1 % yeast extract and 10 μg/mL neomycin sulfate medium may also be used as GAS is intrinsically resistant to this antibiotic. Plates should be dried for 15 min in a sterile environment (such as a sterilized biological safety cabinet) by removing the lid off of the plates and using it as a prop to rest the inverted, agar-laden portion. Once bacteria have been plated dropwise, allow the drops to absorb into the agar before inverting and incubating.

Acknowledgements

This work was supported by operating grants to J.K.M. from the Canadian Institutes of Health Research and the Natural Sciences and Engineering Research Council of Canada.

References

1. Carapetis JR, Steer AC, Mulholland EK, Weber M (2005) The global burden of group A streptococcal diseases. Lancet Infect Dis 5:685–94. doi:10.1016/S1473-3099(05)70267-X
2. Shaikh N, Leonard E, Martin JM (2010) Prevalence of streptococcal pharyngitis and streptococcal carriage in children: a meta-analysis. Pediatrics 126:e557–e564. doi:10.1542/peds.2009-2648
3. Ralph AP, Carapetis JR (2013) Group a streptococcal diseases and their global burden. Curr Top Microbiol Immunol 368:1–27. doi:10.1007/82
4. (1999) Nosocomial group A streptococcal infections associated with asymptomatic health-care workers—Maryland and California, 1997. MMWR Morb Mortal Wkly Rep 48: 163–166
5. Walker MJ, Barnett TC, McArthur JD et al (2014) Disease manifestations and pathogenic mechanisms of group a Streptococcus. Clin Microbiol Rev 27:264–301. doi:10.1128/CMR.00101-13
6. Kasper KJ, Zeppa JJ, Wakabayashi AT et al (2014) Bacterial Superantigens Promote Acute Nasopharyngeal Infection by Streptococcus pyogenes in a Human MHC Class II-Dependent Manner. PLoS Pathog 10, e1004155. doi:10.1371/journal.ppat.1004155
7. Park H, Francis KP, Yu J, Cleary PP (2003) Membranous Cells in Nasal-Associated Lymphoid Tissue: A Portal of Entry for the Respiratory Mucosal Pathogen Group A Streptococcus. J Immunol 171:2532–2537
8. Smoot JC, Barbian KD, Van Gompel JJ et al (2002) Genome sequence and comparative microarray analysis of serotype M18 group A Streptococcus strains associated with acute rheumatic fever outbreaks. Proc Natl Acad Sci U S A 99:4668–4673
9. Sundberg EJ, Li H, Llera AS et al (2002) Structures of two streptococcal superantigens bound to TCR beta chains reveal diversity in the architecture of T cell signaling complexes. Structure 10:687–99
10. Nooh MM, El-Gengehi N, Kansal R et al (2007) HLA transgenic mice provide evidence for a direct and dominant role of HLA class II variation in modulating the severity of streptococcal sepsis. J Immunol 178:3076–3083
11. Alam FM, Bateman C, Turner CE et al (2013) Non-invasive monitoring of Streptococcus pyogenes vaccine efficacy using biophotonic imaging. PLoS One 8, e82123. doi:10.1371/journal.pone.0082123

Chapter 9

Mini-Osmotic Pump Infusion Model to Investigate the Systemic Effects of Chronic Continuous Exposure to Staphylococcal Superantigen in Mice

Ashton L. Krogman, Vaidehi Chowdhary, and Govindarajan Rajagopalan

Abstract

Staphylococcus aureus can exist as a colonizer or can cause a spectrum of diseases. *S. aureus* elaborates several exotoxins and the superantigens are one among them. Staphylococcal superantigens (SSAg) cause robust activation of the immune system and acute exposure to significant amounts of SSAg can be potentially lethal. However, chronic exposure to SSAg is also possible. Administering SSAg using mini-osmotic pumps may mimic chronic recurrent exposure to SSAg. This is a relatively simple and safe way to administer purified SSAg or any other toxin/agent. In this chapter, we describe the mini-osmotic pump-mediated delivery of SSAg.

Key words Mini-osmotic pump, Superantigen, *Staphylococcus aureus*

1 Introduction

Staphylococcus aureus is a successful pathogen. It can asymptomatically colonize 20–30 % of individuals either chronically or intermittently. Skin and the anterior nares are the common sites for colonization [1, 2]. *S. aureus* can also cause a wide spectrum of symptomatic diseases, from localized minor skin infections to life-threatening systemic diseases such as pneumonia, sepsis, and toxic shock syndrome [3].

Staphylococcus aureus produces several exotoxins, which facilitate its existence as a colonizer as well as contribute to its ability to induce highly pathogenic invasive diseases [4, 5]. The staphylococcal superantigens (SSAg) are one among them. SSAg are the most potent biological activators of the immune system known to date [6]. SSAg cause immune activation in a unique manner. SSAg can bind directly to MHC class II molecules expressed on the cell surface of various professional and non-professional antigen-presenting cells without undergoing any classical antigen processing steps.

Amanda J. Brosnahan (ed.), *Superantigens: Methods and Protocols*, Methods in Molecular Biology, vol. 1396,
DOI 10.1007/978-1-4939-3344-0_9, © Springer Science+Business Media New York 2016

By directly binding to MHC class II molecules outside of the peptide-binding groove, SSAg can cross-link MHC class II molecules, activate such SSAg-presenting cells leading to production of several cytokines, chemokines, and other mediators, a majority of which are proinflammatory in nature. MHC class II bound-SSAg can subsequently bind to certain T cell receptor α or β chain variable regions (TCR Vα or β families) irrespective of the antigen specificities of those T cells. This process results in cross-linking of TCRs, rapid and robust activation of such T cells (both $CD4^+$ as well as $CD8^+$), production of various cytokines/chemokines, and mediation of respective effector functions of such activated T cells [7]. For example, activation of Th1 cells by SSAg may lead to production of Th1-type cytokines. Engagement and cross-linking of TCRs on a $CD8^+$ T cell may lead to production of cytokines as well as release of perforins and granzyme, which may mediate cytotoxicity.

Acute exposure to significant amount of SSAg as in pneumonia or sepsis leads to massive and rapid activation of the immune system by the abovementioned processes. This is termed as systemic inflammatory response syndrome and generally culminates in multiple organ dysfunction syndrome, which can be lethal [8]. However, in certain clinical situations, chronic exposure to small amounts of SSAg can occur, e.g., chronic *S. aureus* carriers, chronic or recurrent staphylococcal skin and soft tissue infections, or catheter-associated biofilm infection [9, 10]. The immunopathology ensuing from chronic immune activation mediated by SSAg could be distinct from that resulting from acute exposures. The mini-osmotic pump-mediated systemic delivery of SSAg is an ideal and simple way to investigate this process. The ALZET® Osmotic Pumps are small, infusion pumps commonly used to deliver drugs or agents to unrestrained laboratory animals. We have demonstrated that continuous delivery of small amounts of SSAg using such mini-osmotic pumps causes a distinct systemic inflammatory disease in our humanized mouse model [11]. In this chapter, we describe the protocol to investigate the systemic effects of chronic exposure to SSAg delivered using mini-osmotic pumps.

2 Materials

1. ALZET®mini-osmotic pumps of desired delivery period (1, 3, 7, 14, or 28 days for mice) (DURECT Corporation, Cupertino, CA). Details regarding the functioning of mini-osmotic pumps, choice of the pumps based on the delivery rate, dose, and route of delivery are discussed in detail in the manufacturer's website (http://www.alzet.com/).

2. Staphylococcal superantigen: Either purified SSAg or filtered bacterial culture supernatant containing SSAg, depending on the study question (*see* **Note 1**).
3. 1× Phosphate-buffered saline (PBS): Either prepared from 10× stock or prepared in the lab and filter sterilized.
4. Mice: Age, sex, strain, gene-targeted mice, etc., chosen according to the study question (*see* **Note 2**).
5. Anesthetic agent(s): Preferred by the investigator from the list of agents approved for rodent use as well as approved by Institutional Animal Care and Use Committee (IACUC). The inhalant anesthetic, isoflurane, is ideal for such minor procedures. It is safer, easy to induce, and maintain anesthesia and has a rapid recovery time from anesthesia. However, special equipment and approved laboratory setup to quench isoflurane are needed for safe handling/administration of Isoflurane.
6. Surgical tools: Scissors, scalpel, tissue forceps, curved artery clamps, wound closure stainless steel clips, and applicators; all suitable for rodent surgery and sterilized prior to use.
7. Animal Prep tools: Small clipper, handheld vacuum to remove clippings, topical antiseptics to prepare surgical site, sterile gloves.
8. Post-experimental procedure materials (optional): Serum separation tubes with syringe and needle for blood collection, formalin, Tissue-Tek Optimum Cutting Compound (Sakura FineTek USA INC), RNAlater (Life Technologies).

3 Methods

3.1 Loading Mini-Osmotic Pumps

1. Prepare the SSAg at desired concentration in required volume of PBS corresponding to the capacity of mini-osmotic pumps and number of pumps to be loaded (*see* **Note 3**). (Always prepare the SSAg slightly in excess to adjust for pipetting loss and loss during loading the mini-osmotic pumps using a syringe.)
2. Load the mini-osmotic pumps with appropriate volume of the SSAg solution or PBS alone as directed by the manufacturer using the provided blunt needle.
3. Carefully insert the flow moderator into the sleeve as directed by the manufacturer.
4. Set aside the loaded mini-osmotic pumps in labeled sterile petri dishes and store aseptically till implantation into mice (*see* **Note 4**).

3.2 Surgical Implantation in Mice

1. Induce and maintain anesthesia using appropriate protocol.
2. Remove fur on the flank of anesthetized mice using a clipper.

3. Disinfect surgical site with povidone iodine swab and then wipe with alcohol wipes.
4. For subcutaneous implantation:
   (a) Make a small (1–2 cm) skin incision using surgical scissors.
   (b) Create a subcutaneous pocket using blunt artery forceps, large enough to accommodate the pump.
   (c) Gently insert pump into the subcutaneous pouch with the opening of the moderator away from the incision site.
   (d) Close the skin wound with surgical clips.
5. To place the pumps intra-abdominally:
   (a) Carefully make a small incision on the abdominal muscularis layer, taking care not to injure/damage internal organs or blood vessels.
   (b) Wet the wound site with sterile saline.
   (c) Gently push the pump inside the abdominal cavity, taking care not to damage any tissue/organs.
   (d) Close the abdominal opening with sutures as per IACUC recommended guidelines.
   (e) Close the skin wound with surgical clips.
6. While mice are still under anesthesia, inject saline (to prevent dehydration), analgesics, or antibacterials as approved and mandated by IACUC policies.
7. Return mice to heated blankets to maintain body temperature until animals completely recover from anesthesia.

3.3 Postoperative Follow-Up and Experimental Procedures

1. Monitor the surgical site daily to ensure the presence of intact wound clips and to look for any abnormal wound discharge or infection. Note body weight, animal activity and any abnormal changes in activity/behavior.
2. Remove wound clips 7 days later, if mini-osmotic pumps of greater than 7-day duration are implanted.
3. During the experimental duration, animals can be bled at IACUC recommended intervals to measure serum cytokines or other analyses.
4. At the time of termination of experiment, animals are euthanized as per IACUC guidelines.
5. Immediately after euthanasia, collect blood by cardiac puncture using a syringe and needle in serum separation tubes. Keep the serum tubes refrigerated until sera are separated, aliquoted, and stored frozen at –80°C for measuring cytokine or other immunological/biochemical analyses.

6. Collect spleen, thymus, and lymph nodes and process for flow cytometric analyses or other experiments, accordingly.
7. Collect small pieces of organs in buffered formalin for histopathological analyses or in appropriate fixative for electron micrography.
8. Collect pieces of organs in Tissue-Tek Optimum Cutting Compound for cryopreservation and subsequent immunochemical analyses.
9. Collect organs in RNAlater for subsequent RNA-related assays, if needed.

4 Notes

1. Staphylococcal enterotoxins fall under the Federal Select Agents and Toxins Category. Special permits and institutional biosafety approval are needed to use SSAg. Highly purified, endotoxin-reduced SSAg may be purchased from Toxin Technology Inc, (Sarasota, FL).
2. Appropriate Institutional Animal Care and Use Committee's approval is required for all experiments/procedures involving animals. Special handling procedures should be followed for various chemicals and reagents used including anesthetic agents and SSAg.
3. Other drugs or agents may be mixed with SSAg based on their solubility in PBS and experimental requirement.
4. It is important to note that the mini-osmotic pump will not start releasing SSAg immediately after implantation in mice. This lag period varies depending on the type of pump used. For immediate release of the SSAg from the pumps after implantation, follow the priming steps described in the manufacturer's website depending on the type of osmotic pump used.

Acknowledgements

This work was supported by National Institutes of Health Grant 5K23AR057815-02, an American College of Rheumatology Research and Education Foundation Career Development Bridge Funding Award, and a Ronald F. Kinney Executive Dean for Research Career Development Award from the Mayo Foundation (all to V.R.C.). This study was funded by NIH grants AI101172 and AI68741 (GR).

References

1. Graham PL 3rd, Lin SX, Larson EL (2006) A U.S. population-based survey of *Staphylococcus aureus* colonization. Ann Intern Med 144: 318–325
2. Kluytmans J, van Belkum A, Verbrugh H (1997) Nasal carriage of *Staphylococcus aureus*: epidemiology, underlying mechanisms, and associated risks. Clin Microbiol Rev 10: 505–520
3. DeLeo FR, Otto M, Kreiswirth BN, Chambers HF (2010) Community-associated meticillin-resistant *Staphylococcus aureus*. Lancet 375: 1557–1568
4. Otto M (2014) *Staphylococcus aureus* toxins. Curr Opin Microbiol 17:32–37
5. Foster TJ, Geoghegan JA, Ganesh VK, Hook M (2014) Adhesion, invasion and evasion: the many functions of the surface proteins of *Staphylococcus aureus*. Nat Rev Microbiol 12:49–62
6. Spaulding AR, Salgado-Pabón W, Kohler PL, Horswill AR, Leung DYM, Schlievert PM (2013) Staphylococcal and streptococcal superantigen exotoxins. Clin Microbiol Rev 26:422–447
7. Krakauer T (2013) Update on staphylococcal superantigen-induced signaling pathways and therapeutic interventions. Toxins (Basel) 5:1629
8. Fraser DJ, Proft T (2008) The bacterial superantigen and superantigen-like proteins. Immunol Rev 225:226–243
9. Bachert C, Zhang N, Patou J, van Zele T, Gevaert P (2008) Role of staphylococcal superantigens in upper airway disease. Curr Opin Allergy Clin Immunol 8:34–38
10. Chung J-W, Greenwood-Quaintance KE, Karau MJ, Tilahun A, Khaleghi SR, Chowdhary VR, David CS, Patel R, Rajagopalan G (2015) Superantigens produced by catheter-associated *Staphylococcus aureus* elicit systemic inflammatory disease in the absence of bacteremia. J Leukoc Biol 98(2):271–281, pii: jlb.4A1214-577RR
11. Chowdhary VR, Tilahun AY, Clark CR, Grande JP, Rajagopalan G (2012) Chronic exposure to staphylococcal superantigen elicits a systemic inflammatory disease mimicking lupus. J Immunol 189:2054–2062

Chapter 10

A Sublethal Swine Model for Defining In Vivo Superantigen-Induced Responses Following Exposure to Staphylococcal Enterotoxin B

Kenneth L. Bost, Kenneth J. Piller, Jack Odle, and Chad H. Stahl

Abstract

In vivo responses to bacterially derived superantigen-like toxins have been difficult to define due to the inherent limitations with rodent models and the relevance that the results obtained from such models may, or may not, have for human pathophysiology. Further the use of challenge doses of superantigen toxins that are lethal or supra-lethal complicates analogies to human exposures which are rarely fatal. Here, we utilize the superantigen, staphylococcal enterotoxin B, at doses that are sublethal in a swine model of toxin-induced incapacitation. Relevant dosing using an animal species for which this toxin is a true superantigen distinguishes this model.

Key words In vivo swine model, Superantigen, Sublethal challenge, Staphylococcal enterotoxin B, Cytokine analyses

1 Introduction

Bacterial superantigens [1] are protein exotoxins produced by some human pathogens, most notably *Staphylococcus aureus* and *Streptococcus pyogenes* [2]. Superantigens mediate their pathological effects by cross-linking T cell receptors and MHC class II molecules on antigen-presenting cells outside the antigen-binding site. The result of this nonspecific interaction is to activate a high percentage of T lymphocytes (e.g., 20 %) and the corresponding antigen-presenting cells to release a cytokine cascade which results in a variety of human pathologies whose symptoms vary somewhat with the particular exotoxin. The ability to activate such a large percentage of leukocytes in such a short time frame makes these exotoxins some of the most toxic proteins known to man.

The ability of many superantigen exotoxins to resist destruction by heat, desiccation, proteolysis, and stomach acids is a testament to the inherent stability of these proteins' structure [3, 4].

Amanda J. Brosnahan (ed.), *Superantigens: Methods and Protocols*, Methods in Molecular Biology, vol. 1396,
DOI 10.1007/978-1-4939-3344-0_10, © Springer Science+Business Media New York 2016

This recalcitrance to degradation coupled with their extreme toxicity makes these superantigens well suited for agents of biowarfare or bioterrorism [5, 6]. A renewed interest in the development of vaccines or antagonists of bacterial superantigen exotoxins has occurred as a protection from and a deterrent to such use.

Unfortunately, animal models that have been used in an attempt to define the exotoxin-induced cytokine cascades or the ability of particular vaccine candidates to induce protective immunity suffer from inherent limitations. Taking staphylococcal enterotoxin B (SEB) as an example, potentiating agents have been used to augment toxicity of this exotoxin in rodent models [7–9], sometimes at supra-lethal doses [10]. SEB-induced lethality has been used in piglet [11, 12] and primate models [13, 14]. However the cytokine responses resulting from such high doses of exotoxin exposure do not resemble those obtained when sublethal doses of SEB were used [15]. Based on the usefulness of pigs in agrimedicine [16, 17], we detail SEB-induced incapacitation of swine as a translational model for defining bacterial superantigen pathophysiology and as a tool to evaluate therapeutic interventions.

2 Materials

2.1 Animal Care and Use

1. All procedures using pigs or piglets will need prior approval by the appropriate Animal Care and Use Committee (*see* **Note 1**).
2. Vendors for piglets or pigs will need to be identified that can supply a sufficient number of animals required for the study (e.g., crossbred, mixed gender piglets of desired age) (*see* **Note 2**).
3. Animal housing is dictated by the particular experimental design (*see* **Note 2**). Facilities should adhere to guidelines for the care and use of experimental animals (*see* **Note 1**), as well as the biosafety guidelines for experimental use of toxins (*see* **Note 3**). For protocols which involve insertion of externalized catheters, individual housing of animals is recommended. This allows acclimation of animals to individual pens prior to surgery and limits complications from contact following surgery.
4. Pig diets will vary based on the age of the pig and the experimental design. Commercial diets that meet established nutrient requirements [18] are available for pre-weaning piglets in liquid form (e.g., Ralco Nutrition, Marshall, MN) and as dry diets post-weaning (e.g., Purina Mills, Gray Summit, MO).

2.2 Collecting Physiological Data

1. Body weights require a veterinary scale capable of weighing up to at least 50 kg.
2. Feed amounts and feed weigh back amounts can be recorded daily at each feeding.

3. Core temperatures can be measured using a rectal thermometer. There are also inexpensive implants and wand detectors that can be purchased to record body temperatures (e.g., BioMedic Data Systems, Inc., Seaford, DE).

2.3 Staphylococcal Enterotoxin B

1. Purified staphylococcal enterotoxin B (SEB) can be purchased from several vendors (e.g., Toxin Technology, Sarasota, FL) (*see* **Note 4**).
2. Secure freezer storage is required once recombinant SEB has been received (*see* **Note 3**).
3. Disposable plasticware, gloves, face shields, and lab wear are required for working with such toxins (*see* **Note 3**).
4. Containers for appropriate disposal of SEB or SEB-contaminated materials will also be required (*see* **Note 3**).

2.4 Surgical Supplies

A variety of general anesthesia equipment and surgical instruments are required to perform swine surgeries; a few specific examples are given below (*see* **Note 5**).

1. Telazol®-Ketamine-Xylazine for sedation.
2. Hibiclens scrub and ethanol for sanitization.
3. Bupivacaine: 0.5 %, to be used as a local anesthetic.
4. Materials specifically required for the insertion of central venous catheters (e.g., Arrow International, Inc., Reading, PA) (*see* **Note 2**).
5. Turbigrip® elastic bandage, cast padding, and Vet Wrap for post-surgical site protection.

2.5 Blood Collection and Storage

1. Three milliliter sterile syringes for blood collection from catheterized animals. Alternatively, Vaccutainer™ tubes (Becton Dickinson) can be used (*see* **Note 6**).
2. Tubes to transfer blood from syringes for centrifugation (*see* **Note 6**).
3. A clinical centrifuge with appropriate rotor and tube holders capable of separating plasma or serum (*see* **Note** 7).
4. Pipetters to remove serum or plasma from tubes following centrifugation.
5. Freezer or refrigerator and tubes for storing aliquots of serum or plasma at −20 °C or 4 °C, respectively (*see* **Note 8**).

2.6 Assays for Inflammatory Mediators

1. Pipetters for dispensing and diluting aliquots of plasma or serum.
2. ELISA antibody pairs or ELISA kits to quantify inflammatory mediators in serum or plasma (*see* **Note 9**). Alternatively, cytometric bead or Luminex-based assays have been recently developed (*see* **Note 10**).
3. Equipment required to detect and quantify the assay system that is selected (*see* **Note 10**).

2.7 Euthanasia of Swine following Data Collection

1. Pentobarbital for euthanasia following data collection in this sublethal model (*see* **Notes 1**, **4**, and **5**). An alternative method for euthanasia uses a penetrating captive bolt if there is a need to not have tissues exposed to pentobarbital.
2. Appropriate containers for carcass disposal (*see* **Notes 1**, **4**, and **5**).

3 Methods

3.1 Animal Housing

Animal housing is dictated by the particular experimental design (*see* **Note 1**).

1. A minimum of 8 square feet per animal up to a body weight of 15 kg, and 24 square feet up to a body weight of 100 kg, is required.
2. A 15-h light and 9-h dark cycle with ad libitum access to water and a typical nursery pig diet are standard operating procedures.
3. Housing at a thermal neutral temperature of approximately 28 °C until reaching a body weight of 20 kg, approximately 24 °C until reaching a body weight of 85 kg, and then approximately 18 °C when exceeding a body weight of 100 kg is recommended.
4. For protocols which involve insertion of externalized catheters, individual housing of animals is recommended. This allows acclimation of animals to individual pens prior to surgery and limits complications from contact following surgery. Typically, pigs are allowed to acclimate to their new surroundings for approximately 1 week and to allow time to establish baseline physiological measurements.

3.2 Monitoring Food Consumption, Body Weight, and Core Temperature

It is best to record these parameters for the first week prior to beginning the experimental protocol to establish a baseline for food consumption and growth rate.

1. Body weight can be recorded daily by weighing using appropriate capacity scales.
2. Food consumption is most easily monitored by housing pigs individually. Feed amounts and feed weigh back amounts can be recorded at each feeding.
3. Core temperatures can be monitored using a rectal thermometer or using wireless implanted devices.

3.3 Insertion of Intravenous Catheters

To facilitate toxin administration, and to allow rapid and frequent blood collection, intravenous catheters can be surgically implanted [15] (*see* **Note 5**). There are several methods which can be used, and the following procedure is a typical one for subcutaneous catheter insertion [19].

1. Pigs should be fasted for 12 h prior to sedation with an intramuscular injection of Telazol®-Ketamine-Xylazine at a dose of 0.02 mL/kg bodyweight, with additional dosing given to effect.
2. Following sedation, pigs are placed in dorsal recumbence and the lateral sides of the neck shaved and aseptically prepared with Hibiclens scrub and ethanol.
3. Prior to initiation of the catheterization with the introducer needle, pigs should be given a subcutaneous infusion of bupivacaine (0.5%, 1.0 mL) approximately 1 to 2 cm caudal of the catheter insertion.
4. Catheterizations on pigs can be performed using 18-gauge, 20 cm single-lumen central venous catheter kits (*see* **Note 2**). Catheterization is initiated with the insertion of a 6.4 cm 18-gauge introducer needle to puncture the external jugular vein.
5. After venipuncture, a flexible guide wire (j-wire length, 45 cm) is inserted into the lumen of the external jugular vein via the introducer needle after which the introducer needle is withdrawn.
6. A rigid vessel dilator is then threaded over the guide wire, advanced through the skin and subcutaneous tissues, and withdrawn.
7. A single-lumen central venous catheter is filled with heparinized saline (10 U/mL) and advanced over the guide wire into the vessel to approximately the level of the cranial right atrium. Once the catheter is in place, the guide wire is removed and the catheter flushed with heparinized saline to confirm patency.
8. A luer-lock catheter cap is placed on the catheter hub, with the catheter being secured to the skin using an Integral Suture Wing.
9. The neck and upper torso of the animals are first protected with Tubigrip® elastic bandage, and then wrapped using cast padding and Vet Wrap.
10. Pigs are allowed to recover in a pen and are monitored for respiration and activity every 15 min until they are sternal, and then every 60 min until they are standing and drinking or eating. Postoperative pain should be assessed by applying slight pressure to the catheterization site. Continuation of an opiate or nonsteroidal anti-inflammatory analgesic may be instituted as needed or as required.
11. Several days following this surgery, a pre-toxin challenge blood sample should be taken, and baseline physiological parameters established for each pig. During the course of the experiment, the inserted catheter allows repeated blood samples to be drawn. Based on our previous experience [15], it is possible to

take multiple 1.0 to 2.0 ml blood samples (e.g., 0, 1, 2, 4, 8 h) following toxin exposures. This provides up to 1.0 ml of serum for each time point for each pig. Larger amounts of blood can be taken depending on animal size.

3.4 Exposure of Pigs to SEB

There are several important considerations for the experimental design including the method of toxin exposure (*see* **Note 11**) and a determination of the dose of SEB (*see* **Note 12**) to be used.

1. For a typical intravenous exposure, a dose of SEB is diluted in saline and administered as a 0.5 ml push via the indwelling catheter.
2. An additional 1.0 ml of saline is then used to flush the catheter and assure that no SEB remains in the catheter or pressure hub.
3. We have found that intravenous doses of SEB between 5 and 25 μg/kg produced at least one external symptom of incapacitation in all exposed pigs without causing any fatalities. Such external symptomology could include prostration, lethargy, emesis, and/or diarrhea.
4. Control animals should receive the same volume of vehicle with a similar amount of an irrelevant protein (e.g., pig serum albumin). We have observed no discernable clinical effects in such control animals.

3.5 Blood Collection

1. Immediately prior to SEB exposure (time 0), a blood sample should be collected via the indwelling catheter, and then at prescribed times following toxin challenge (e.g., 1, 2, 4, 8 h). Collection of blood into coagulation or non-coagulation tubes allows the isolation of serum or plasma, respectively (*see* **Notes 6** and 7).
2. To isolate serum, clotting is allowed to occur at room temperature for approximately 1 h. Tubes can then be placed at 4 °C for 1–18 h to allow clots to contract before centrifugation (e.g., 2000 × *g* for 10 min, *see* **Note** 7). Serum can then be pipetted, and assayed immediately, or stored frozen in aliquots for future analyses (*see* **Note 8**).
3. Plasma can also be isolated by centrifugation (e.g., 2000 × *g* for 20 min, *see* **Note** 7) immediately following collection. Assays can be performed immediately or plasma can be stored at 4 °C for future analyses (*see* **Note 8**).

3.6 Quantification of Serum Cytokine and Acute-Phase Protein following SEB Challenge

1. The amount of serum or plasma obtained from each animal at each time point (~1.0 ml) allows numerous parameters to be measured from a single sample.
2. ELISA reagents or kits can be used to quantify the levels of inflammatory mediators from pig serum or plasma (*see* **Note 9**).

3. Alternative to standard ELISAs, inflammatory parameters can be followed using more recently developed cytometric bead assays [20] or Luminex™-based assays [21] (*see* **Note 10**).
4. Follow the manufacturer's established protocols for assaying these parameters in obtained samples.

4 Notes

1. Guidelines for the care and use of laboratory animals must be followed in all animal housing, procedures, and carcass disposal [22].
2. Some factors that need to be considered when developing the experimental design include the use of inbred versus outbred animals, their age and weight, the group size required for statistical significance, the animals' diet, and their gender. For example, pig or piglet weight will dictate the total amount of SEB that will be required for each animal and for each experiment (*see also* **Note 12**). Furthermore, pig or piglet age and weight will dictate the catheter size to be used and its placement distance from the heart. While an 18-gauge, 20 cm catheter is used as an example, piglets would require a smaller gauge catheter kit.
3. Working in the laboratory with SEB and its use in animals will also need prior approval from the Biosafety and Animal Care and Use Committees. Furthermore, standard operating procedures for storage, handling, and disposal of SEB or SEB-contaminated materials will be required [23]. For example, all disposable materials that come in contact with SEB should be inactivated with 10 % bleach for 30 min before autoclaving or disposal.
4. Since SEB is designated as a biohazard and select agent, purchasing, shipping, and receiving this exotoxin require prior approvals from the appropriate Biosafety Committee. In addition, there are limits as to the amount of SEB that can be purchased and limits for amounts that can be stored once received. These considerations should be factored into the experimental design.
5. The following reference is an excellent resource when designing experiments that involve general anesthesia and surgery in swine [24].
6. Blood collection from catheterized pigs can be accomplished using syringes or Vaccutainer™ tubes. Advantages of syringes include their cost, ease of use, and the ability to pretreat with the anticoagulant of choice. Disadvantages include the necessity to transfer blood samples from syringes into appropriate collection tubes prior to centrifugation. Vaccutainer™ tubes

eliminate the need for transferring blood samples and tubes can be purchased for collecting plasma that already contain anticoagulant.

7. The type of collection tube utilized will depend on the assays to be performed. Most ELISAs work best with serum; however some assays require plasma for quantification of serum parameters. For isolation of serum, blood should be allowed to clot at room temperature for approximately 1 h. Tubes can then be placed at 4 °C for 1 to 18 h to allow clots to contract before centrifugation. Centrifugation at 2000 × *g* for 10 min in a refrigerated centrifuge will allow separation of serum from the clot. For isolation of plasma, blood should be collected with the anticoagulant of choice. Centrifugation at 2000 × *g* for 20 min separates cells and platelets from the plasma. It is important to know the centrifugal force that the coagulation or non-coagulation tubes can withstand prior to centrifugation to reduce the risk of tubes breaking or deforming.
8. Serum or plasma can be stored for periods of time prior to performing analyses. Serum can be stored in non-frost-free freezers (−20 °C to −80 °C) for extended periods of time (e.g., 1–8 months) prior to analyses without significant reduction in many inflammatory mediators. Note however that some inflammatory mediators may be best quantified with freshly isolated serum. Plasma can be stored at 4 °C for a limited time (e.g., 1 week) prior to analyses for many inflammatory mediators. Note however that some inflammatory mediators may be best quantified with freshly isolated plasma. It is not recommended to freeze aliquots of plasma as this process may result in the clotting of some samples, making analyses difficult or impossible.
9. There are numerous companies which sell pairs of antibodies (e.g., Porcine DuoSet, R&D Systems, Minneapolis, MN) or ELISA kits (e.g., RayBiotech, Norcross, GA) to quantify a variety of cytokine and acute-phase protein levels. Such assays typically have visible color change as a readout which can be quantified using standard ELISA plate readers (e.g., Cole-Palmer Bio-Tek® ELX 800™).
10. More recent alternatives to standard ELISAs for quantifying inflammatory mediators include cytometric bead assays or Luminex™-based assays which are also commercially available (e.g., Millipore, MILLIPLEX 13-plex Porcine Cytokine/Chemokine Panel). Prior to purchasing commercially available reagents to quantify inflammatory mediators in pig serum or plasma samples, it is important to identify the equipment that will be required to detect the signal generated. Typical ELISAs require a microplate reader that can quantify absorbance of visible color change while other assay systems may require fluorescent signals to be detected and quantified.

11. One important consideration for the experimental design is to determine the most appropriate method for toxin exposure (e.g., oral, intranasal, intramuscular, aerosol). The method to be utilized is often dictated by the scenario being modeled. For example, studies investigating food contamination would require oral administration. Modeling an aerosol exposure can be technically challenging as aerosol generators or chambers are required. Further, the generation of SEB-containing aerosols is a significant biosafety concern.
12. A second consideration is the sublethal dose of SEB to be used. We have experienced varying degrees of incapacitation with equivalent dosing from different lots or different commercial vendors supplying purified SEB. If possible, a single lot of purified SEB should be used within the context of an individual experiment.

References

1. Marrack P, Kappler J (1990) The staphylococcal enterotoxins and their relatives. Science 248:705–711
2. Spaulding AR, Salgado-Pabon W, Kohler PL, Horswill AR, Leung DY et al (2013) Staphylococcal and streptococcal superantigen exotoxins. Clin Microbiol Rev 26:422–447
3. Papageorgiou AC, Tranter HS, Acharya KR (1998) Crystal structure of microbial superantigen staphylococcal enterotoxin B at 1.5 A resolution: implications for superantigen recognition by MHC class II molecules and T-cell receptors. J Mol Biol 277:61–79
4. Swaminathan S, Furey W, Pletcher J, Sax M (1992) Crystal structure of staphylococcal enterotoxin B, a superantigen. Nature 359: 801–806
5. Christopher GW, Cieslak TJ, Pavlin JA, Eitzen EM Jr (1997) Biological warfare. A historical perspective. JAMA 278:412–417
6. Madsen JM (2001) Toxins as weapons of mass destruction. A comparison and contrast with biological-warfare and chemical-warfare agents. Clin Lab Med 21:593–605
7. Boles JW, Pitt ML, LeClaire RD, Gibbs PH, Torres E et al (2003) Generation of protective immunity by inactivated recombinant staphylococcal enterotoxin B vaccine in nonhuman primates and identification of correlates of immunity. Clin Immunol 108:51–59
8. Faulkner L, Cooper A, Fantino C, Altmann DM, Sriskandan S (2005) The mechanism of superantigen-mediated toxic shock: not a simple Th1 cytokine storm. J Immunol 175: 6870–6877
9. Stiles BG, Garza AR, Ulrich RG, Boles JW (2001) Mucosal vaccination with recombinantly attenuated staphylococcal enterotoxin B and protection in a murine model. Infect Immun 69:2031–2036
10. Ulrich RG, Olson MA, Bavari S (1998) Development of engineered vaccines effective against structurally related bacterial superantigens. Vaccine 16:1857–1864
11. Bi S, Das R, Zelazowska E, Mani S, Neill R et al (2009) The cellular and molecular immune response of the weanling piglet to staphylococcal enterotoxin B. Exp Biol Med (Maywood) 234:1305–1315
12. van Gessel YA, Mani S, Bi S, Hammamieh R, Shupp JW et al (2004) Functional piglet model for the clinical syndrome and postmortem findings induced by staphylococcal enterotoxin B. Exp Biol Med (Maywood) 229:1061–1071
13. Komisar JL, Weng CF, Oyejide A, Hunt RE, Briscoe C et al (2001) Cellular and cytokine responses in the circulation and tissue reactions in the lung of rhesus monkeys (Macaca mulatta) pretreated with cyclosporin A and challenged with staphylococcal enterotoxin B. Toxicol Pathol 29:369–378
14. Weng CF, Komisar JL, Hunt RE, Johnson AJ, Pitt ML et al (1997) Immediate responses of leukocytes, cytokines and glucocorticoid hormones in the blood circulation of monkeys following challenge with aerosolized staphylococcal enterotoxin B. Int Immunol 9: 1825–1836
15. Hudson LC, Seabolt BS, Odle J, Bost KL, Stahl CH et al (2013) Sublethal staphylococcal

enterotoxin B challenge model in pigs to evaluate protection following immunization with a soybean-derived vaccine. Clin Vaccine Immunol 20:24–32
16. Odle J, Lin X, Jacobi SK, Kim SW, Stahl CH (2014) The Suckling Piglet as an Agrimedical Model for the Study of Pediatric Nutrition and Metabolism. Ann Rev Anim Biosci 2:419–434
17. Swindle MM, Makin A, Herron AJ, Clubb FJ Jr, Frazier KS (2012) Swine as models in biomedical research and toxicology testing. Vet Pathol 49:344–356
18. Council NR (2012) Nutrition requirements of swine. National Academies Press, Washington, DC
19. Flournoy WS, Mani S (2009) Percutaneous external jugular vein catheterization in piglets using a triangulation technique. Lab Anim 43:344–349
20. Wyns H, Croubels S, Demeyere K, Watteyn A, De Backer P et al (2013) Development of a cytometric bead array screening tool for the simultaneous detection of pro-inflammatory cytokines in porcine plasma. Vet Immunol Immunopathol 151:28–36
21. Lawson S, Lunney J, Zuckermann F, Osorio F, Nelson E et al (2010) Development of an 8-plex Luminex assay to detect swine cytokines for vaccine development: assessment of immunity after porcine reproductive and respiratory syndrome virus (PRRSV) vaccination. Vaccine 28:5356–5364
22. Worlein JM, Baker K, Bloomsmith M, Coleman K, Koban TL (2011) The Eighth Edition of the Guide for the Care and Use of Laboratory Animals (2011); Implications for Behavioral Management. Am J Primatol 73:98–98
23. Chosewood LC, Wilson DE (2009) Biosafety in microbiological and biomedical laboratories. U.S. Dept. of Health and Human Services, Washington, DC, pp. xxii-415
24. Swindle MM (2007) Swine in the laboratory: surgery, anesthesia, imaging, and experimental techniques. CRC Press, Boca Raton, FL

Chapter 11

Monkey Feeding Assay for Testing Emetic Activity of Staphylococcal Enterotoxin

Keun Seok Seo

Abstract

Staphylococcal enterotoxins (SEs) are unique bacterial toxins that cause gastrointestinal toxicity as well as superantigenic activity. Since systemic administration of SEs induces superantigenic activity leading to toxic shock syndrome that may mimic enterotoxic activity of SEs such as vomiting and diarrhea, oral administration of SEs in the monkey feeding assay is considered as a standard method to evaluate emetic activity of SEs. This chapter summarizes and discusses practical considerations of the monkey feeding assay used in studies characterizing classical and newly identified SEs.

Key words Staphylococcal enterotoxin, Emesis, Monkey feeding assay

1 Introduction

Staphylococcal food poisoning (SFP) is usually a self-limiting disease with common symptoms including abdominal/muscle cramps, diarrhea, headaches, and emesis. Symptoms typically develop within 4–6 h after ingestion of contaminated food and last for 1–88 h. The first report that staphylococci produce an enterotoxin causing emesis and diarrhea was by Dack et al. [1]. They isolated a large number of pigment-forming staphylococci from a Christmas cake responsible for illness with vomiting and diarrhea in several individuals, and demonstrated that a culture filtrate of the organism reproduced illness when ingested by human volunteers. To date, 23 different staphylococcal enterotoxins (SEs) and staphylococcal enterotoxin-like toxins (SEls) have been identified [2–7] and named as recommended by the International Nomenclature Committee for Staphylococcal Superantigens (INCSS) [8]. The INCSS recommends that to be classified as SEs, the toxins exhibit emetic activity in monkey feeding assay and SEls are defined as toxins that did not exhibit emetic activity in monkey feeding assay or have not been tested yet. According to this guideline, SEs include SEA-SEE, SEG-SEI, and SEK-SET that

Amanda J. Brosnahan (ed.), *Superantigens: Methods and Protocols*, Methods in Molecular Biology, vol. 1396,
DOI 10.1007/978-1-4939-3344-0_11,

demonstrated emetic activity in the monkey feeding assay [4–6, 9, 10]. SEls include SElU-SElX that have not been tested yet. Results of monkey feeding assays are summarized in Table 1.

Although the specific cell types or the receptors in abdomen that interact with SEs and cause emesis have not been clearly identified, studies in primates demonstrated that SEs stimulate local neural receptors in abdomen which transmit impulses to the medullary emetic center at the fourth ventricle in the brain via vagus and sympathetic nerves [11]. Other studies suggested the involvement of inflammatory mediators and neural transmitters in the SE-induced emetic response. These studies demonstrated that oral administration of SEs induced elevated levels of vasoactive inflammatory mediators such as prostaglandin E_2, leukotriene B4, 5-hydroxyeicosatetraenoic acid, and release of substance P by mast cells which may trigger the emetic center in the brain [11–13].

Toxic shock syndrome (TSS) is caused by superantigenic activity of SEs that induce massive activation T cells with uncontrolled production of cytokines, resulting in shock and fever. Some TSS patients showed similar gastrointestinal symptoms to those seen in SFP patients including vomiting and diarrhea. This raises the question of whether superantigenicity of SEs might be related to emetic activity of SEs. However, several pieces of evidence suggest that superantigenicity of SEs is not directly responsible for emetic activity of SEs. Mutagenesis of several SEs revealed that a domain responsible for T cell activation is separated from a domain responsible for emetic activity of SEs. These studies demonstrated that SEA, SEB, and SEC mutant toxins lacking superantigenicity still retain ability to induce emesis and vice versa [14–16]. Additionally, SFP patients rarely showed TSS like symptoms such as shock and fever. Considering that systemic exposures to SEs cause TSS, which may also mimic enterotoxic activity of SEs, oral administration of SEs is the preferred method to evaluate enterotoxic activity of SEs. The monkey feeding assay developed by Bergdoll [9] is considered to be the standard method for determining the emetic activity of SEs following oral administration. Other animal models including dog, piglet, and house musk shrews have also been developed [17–19]. However, results with these animal models requiring systemic administration of SEs need to be carefully interpreted, since they may mimic TSS rather than SFP.

2 Materials

2.1 Preparation of SEs

In the literature, SEs used in the monkey feeding assay have mainly been prepared by two different methods. Early studies have used SEs expressed in non-toxigenic host, *Staphylococcus aureus* RN4220 strain harboring a single SE gene cloned into a plasmid. Most recent studies have used SEs expressed in *Escherichia coli* harboring

Table 1
Emetic activity of SEs in monkey feeding assay

Toxin	Animal	Administration	Toxin preparation	Dose	No. of tested	No. of vomiting	Reference
SEA	*Macaca mulatta*	Nasogastric tube	Purified toxin expressed in *S. aureus*	5 μg/animal	10	4	[9]
	Macaca mulatta	Nasogastric tube	Purified toxin expressed in *S. aureus*	10 μg/animal	30	16	[9]
	Macaca mulatta	Nasogastric tube	Purified toxin expressed in *S. aureus*	20 μg/animal	12	9	[9]
	Macaca fascicularis	Nasogastric tube	Purified toxin expressed in *S. aureus*	10 μg/kg	10	5	[10]
	Macaca fascicularis	Nasogastric tube	Purified toxin expressed in *E. coli*	100 μg/kg	7	6	[10]
SEB	*Macaca mulatta*	Nasogastric tube	Purified toxin expressed in *S. aureus*	12 μg/animal	16	9	[9]
SEC	*Macaca mulatta*	Nasogastric tube	Purified toxin expressed in *S. aureus*	3 μg/animal	6	5	[9]
SED	*Macaca mulatta*	Nasogastric tube	Purified toxin expressed in *S. aureus*	10–20 μg/animal[a]	NA[b]	NA	[9]
SEE	NT[c]						
SEG	*Macaca mulatta*	Nasogastric tube	Purified toxin expressed in *S. aureus*	80 μg/kg	6	4	[4]
SEH	*Macaca mulatta*	Nasogastric tube	Purified toxin expressed in *S. aureus*	30 μg/kg	1	1	[6]
SEI	*Macaca mulatta*	Nasogastric tube	Purified toxin expressed in *S. aureus*	150 μg/kg	4	1	[4]
SElJ	NT						
SEK	*Macaca fascicularis*	Nasogastric tube	Purified toxin expressed in *E. coli*	100 μg/kg	6	2	[10]

(continued)

Table 1 (continued)

Toxin	Animal	Administration	Toxin preparation	Dose	No. of tested	No. of vomiting	Reference
SEL	*Macaca fascicularis*	Nasogastric tube	Purified toxin expressed in *E. coli*	100 μg/kg	6	1	[10]
SEM	*Macaca fascicularis*	Nasogastric tube	Purified toxin expressed in *E. coli*	100 μg/kg	7	1	[10]
SEN	*Macaca fascicularis*	Nasogastric tube	Purified toxin expressed in *E. coli*	100 μg/kg	6	2	[10]
SEO	*Macaca fascicularis*	Nasogastric tube	Purified toxin expressed in *E. coli*	100 μg/kg	8	1	[10]
SEP	*Macaca fascicularis*	Nasogastric tube	Purified toxin expressed in *E. coli*	100 μg/kg	6	3	[10]
SEQ	*Macaca fascicularis*	Nasogastric tube	Purified toxin expressed in *E. coli*	100 μg/kg	6	2	[10]
SER	*Macaca fascicularis*	Nasogastric tube	Purified toxin expressed in *E. coli*	100 μg/kg	4	2	[5]
SES	*Macaca fascicularis*	Nasogastric tube	Purified toxin expressed in *E. coli*	100 μg/kg	4	2	[5]
SET	*Macaca fascicularis*	Nasogastric tube	Purified toxin expressed in *E. coli*	100 μg/kg	4	3	[5]
SElU	NT						
SElV	NT						
SElX	NT						

[a]The number shown is ED50 (a dose-induced 50 % animals were emetic) of SED
[b]Not available
[c]Not tested

a single SE gene cloned into a plasmid with inducible promoter and tagging system. The use of SEs expressed in *S. aureus* is preferred since it warrants physicochemical properties of SEs as their native forms. However, it is relatively difficult to purify SEs in high purity from *S. aureus*, as compared to other methods, primarily due to contamination by other toxins such as hemolysins and leukotoxins which affect host responses to SEs [20, 21]. The use of SEs expressed in *E. coli* excludes the possibility of contamination of other toxins and is preferred due to the simplicity of purification and the large amount of protein that can be produced. However, contamination by endotoxin may become problematic since SEs are able to enhance endotoxin-mediated shock responses including vomiting and diarrhea [22]. Therefore, it is essential to ensure the removal of endotoxin in the SE preparation. Detailed SE preparation methods are described in Chapter 2.

2.2 Animals

Male and female animals ranging 2–5 kg in weight have been used in the monkey feeding assay. Animals are individually housed at 23–27 °C and 50–70 % humidity and a 12-h/12-h light/dark cycle. The rhesus monkey (*Macaca mulatta*) have been used in the method developed by Bergdoll to characterize emetic activity of classical SEs including SEA-SEI [4, 6, 9]. The cynomolgus monkey (*Macaca fascicularis*) was used in the study to characterize emetic activity of newly discovered SEs including SEK-SET [10].

3 Methods

1. Fast animals for 24 h before administering SEs.
2. While one animal handler holds the animal from the back between the knees with one hand holding the head and the other hand holding the mouth open, the other animal handler inserts a nasogastric tube (outside diameter 5 mm, bore 3 mm) lubricated with glycerol directly to the esophagus until it reaches to the stomach (*see* **Note 1**).
3. Purified SE dissolved in 50 ml of NZ-Amine NAK nutrient broth or 10 ml of distilled water is delivered through a nasogastric tube using a syringe.
4. After administering SE, animals are monitored for emesis up to 5 h. The number of vomiting events, the time to the vomiting event (latency period), and behavioral changes are recorded. Occasionally, an animal may vomit due to the intubation procedures and other provocation. It may require at least two vomiting events to be positive reaction.

4 Notes

1. For the safety of animal handler, animals may be sedated with ketamine-HCl. A study found that the rhesus monkey (*Macaca mulatta*) did not show emetic response to intramuscular injection of ketamine-HCl at doses of 7.5–15 mg/kg, and sedation did not inhibit emetic activity of subsequent oral administration of SEB [23]. However, another other study demonstrated that 10 of 41 cynomolgus monkey (*Macaca fascicularis*) showed emetic response to 2.5–20 mg/kg of ketamine-HCl intramuscular injection [24].

References

1. Dack GM, Cary WE, Woolper O, Wiggers H (1930) An outbreak of food poisoning proved to be due to a yellow hemolytic *Staphylococcus*. Can J Microbiol 4:167–175
2. Bergdoll MS, Surgalla MJ, Dack GM (1959) Staphylococcal enterotoxin. Identification of a specific precipitating antibody with enterotoxin-neutralizing property. J Immunol 83:334–338
3. Casman EP (1960) Further serological studies of staphylococcal enterotoxin. J Bacteriol 79: 849–856
4. Munson SH, Tremaine MT, Betley MJ, Welch RA (1998) Identification and characterization of staphylococcal enterotoxin types G and I from *Staphylococcus aureus*. Infect Immun 66:3337–3348
5. Ono HK, Omoe K, Imanishi K, Iwakabe Y, Hu DL, Kato H, Saito N, Nakane A, Uchiyama T, Shinagawa K (2008) Identification and characterization of two novel staphylococcal enterotoxins, types S and T. Infect Immun 76:4999–5005. doi:10.1128/iai.00045-08
6. Su YC, Wong AC (1995) Identification and purification of a new staphylococcal enterotoxin, H. Appl Environ Microbiol 61: 1438–1443
7. Spaulding AR, Salgado-Pabon W, Kohler PL, Horswill AR, Leung DY, Schlievert PM (2013) Staphylococcal and streptococcal superantigen exotoxins. Clin Microbiol Rev 26:422–447. doi:10.1128/cmr.00104-12
8. Lina G, Bohach GA, Nair SP, Hiramatsu K, Jouvin-Marche E, Mariuzza R (2004) Standard nomenclature for the superantigens expressed by *Staphylococcus*. J Infect Dis 189:2334–2336. doi:10.1086/420852
9. Bergdoll MS (1988) Monkey feeding test for staphylococcal enterotoxin. Methods Enzymol 165:324–333
10. Omoe K, Hu DL, Ono HK, Shimizu S, Takahashi-Omoe H, Nakane A, Uchiyama T, Shinagawa K, Imanishi K (2013) Emetic potentials of newly identified staphylococcal enterotoxin-like toxins. Infect Immun 81:3627–3631. doi:10.1128/iai.00550-13
11. Sugiyama H, Hayama T (1965) Abdominal viscera as site of emetic action for staphylococcal enterotoxin in the monkey. J Infect Dis 115:330–336
12. Alber G, Scheuber PH, Reck B, Sailer-Kramer B, Hartmann A, Hammer DK (1989) Role of substance P in immediate-type skin reactions induced by staphylococcal enterotoxin B in unsensitized monkeys. J Allergy Clin Immunol 84:880–885
13. Jett M, Neill R, Welch C, Boyle T, Bernton E, Hoover D, Lowell G, Hunt RE, Chatterjee S, Gemski P (1994) Identification of staphylococcal enterotoxin B sequences important for induction of lymphocyte proliferation by using synthetic peptide fragments of the toxin. Infect Immun 62:3408–3415
14. Alber G, Hammer DK, Fleischer B (1990) Relationship between enterotoxic- and T lymphocyte-stimulating activity of staphylococcal enterotoxin B. J Immunol 144:4501–4506
15. Harris TO, Betley MJ (1995) Biological activities of staphylococcal enterotoxin type A mutants with N-terminal substitutions. Infect Immun 63:2133–2140
16. Hovde CJ, Marr JC, Hoffmann ML, Hackett SP, Chi YI, Crum KK, Stevens DL, Stauffacher CV, Bohach GA (1994) Investigation of the role of the disulphide bond in the activity and structure of staphylococcal enterotoxin C1. Mol Microbiol 13:897–909
17. van Gessel YA, Mani S, Bi S, Hammamieh R, Shupp JW, Das R, Coleman GD, Jett M (2004) Functional piglet model for the clinical syndrome and postmortem findings induced by staphylococcal enterotoxin B. Exp Biol Med 229:1061–1071

18. Hu DL, Omoe K, Saleh MH, Ono K, Sugii S, Nakane A, Shinagawa K (2001) Analysis of the epitopes on staphylococcal enterotoxin A responsible for emetic activity. J Vet Med Sci 63:237–241
19. Kocandrle V, Houttuin E, Prohaska JV (1966) Acute hemodynamic and gastrointestinal changes produced by staphylococcal exotoxin and enterotoxin in dogs. J Surg Res 6:50–57
20. Onogawa T (2002) Staphylococcal alpha-toxin synergistically enhances inflammation caused by bacterial components. FEMS Immunol Med Microbiol 33:15–21
21. Holzinger D, Gieldon L, Mysore V, Nippe N, Taxman DJ, Duncan JA, Broglie PM, Marketon K, Austermann J, Vogl T, Foell D, Niemann S, Peters G, Roth J, Loffler B (2012) *Staphylococcus aureus* Panton-Valentine leukocidin induces an inflammatory response in human phagocytes via the NLRP3 inflammasome. J Leukoc Biol 92:1069–1081. doi:10.1189/jlb.0112014
22. Kulhankova K, King J, Salgado-Pabon W (2014) Staphylococcal toxic shock syndrome: superantigen-mediated enhancement of endotoxin shock and adaptive immune suppression. Immunol Res 59:182–187. doi:10.1007/s12026-014-8538-8
23. Melling J (1977) Use of anaesthesia with ketamine HCl in assessing the emetic activity of staphylococcal enterotoxin B in the rhesus monkey (*Macacca mulatta*). Br J Exp Pathol 58:40–41
24. Adesiyun AA, Tatini SR (1982) Incidence of ketamine-induced emesis in cynomologus monkeys (*Macaca fascicularis*) used for staphylococcal enterotoxin bioassay. Br J Exp Pathol 63:330–335

Chapter 12

Use of an Ex Vivo Porcine Mucosal Model to Study Superantigen Penetration

Christopher A. Squier and Mary J. Mantz

Abstract

In vitro perfusion studies are frequently used to determine the penetration of compounds through skin and mucosa. Porcine tissue has been shown to be an excellent model for human tissue in terms of structure, function, and reactivity. We describe the use of porcine tissue ex-vivo in a continuous flow perfusion system to study the behavior of superantigens in this model.

Key words Permeability, Pathogenic microorganisms, Superantigens, Perfusion cells, Mucosa, Permeability barrier, Ex-vivo porcine model

1 Introduction

Protection of tissues and organs from external pathogenic organisms and their toxins, including superantigens, depends on the barrier properties of the covering and lining tissues of the body. The skin represents the principal protective layer of the outer surface of the body and consists of a waterproof and tough cornified layer comprising a number of layers of stratified and differentiating cells termed a stratified squamous epithelium. The tissues lining the various cavities which communicate with the exterior of the body, such as the mouth and alimentary tract and the genitourinary passages, are termed mucosae and consist of tissues with a variety of structural organization. This ranges from the stratified squamous epithelium which lines the mouth, esophagus, vagina, and uterine cervix to simple epithelium, consisting of a single layer of cells, lining the whole of the digestive tract between the stomach and the anus. Staphylococci and their toxins have been described on skin and in the oral cavity and vagina [1–3]. In this chapter we focus on models for studying antigen penetration across skin, the vagina, and the oral cavity.

Amanda J. Brosnahan (ed.), *Superantigens: Methods and Protocols*, Methods in Molecular Biology, vol. 1396, DOI 10.1007/978-1-4939-3344-0_12, © Springer Science+Business Media New York 2016

Norris [4], discussing the pathogenic role of superantigens on the skin, noted that "without penetration there is nothing." The barrier properties of epithelia are thus paramount in determining whether there will be local or systemic disease as a result of antigenic challenge. For skin and oral, vaginal, and cervical mucosa the principal barrier to the penetration of relatively large, hydrophilic compounds such as superantigens is located in the intercellular regions between the surface cells of the tissue [5]. In keratinized epithelia such as that in skin and the oral gingiva and palate, small intracellular organelles become evident in the epithelial prickle cell layer (*see* Fig. 1) [6]. These granules contain lipid organized into parallel lamellae and are known as membrane-coating granules or lamellar granules. At the boundary between the granular and keratinized layers, the membrane-coating granules migrate to the superficial aspect of the epithelial cell, where the bounding membrane of the granule fuses with the cell plasma membrane so that the lipid lamellae are extruded into the extracellular spaces of the surface layer [7]. The barrier properties of the stratum corneum are related to the phase behavior of the intercellular lipids.

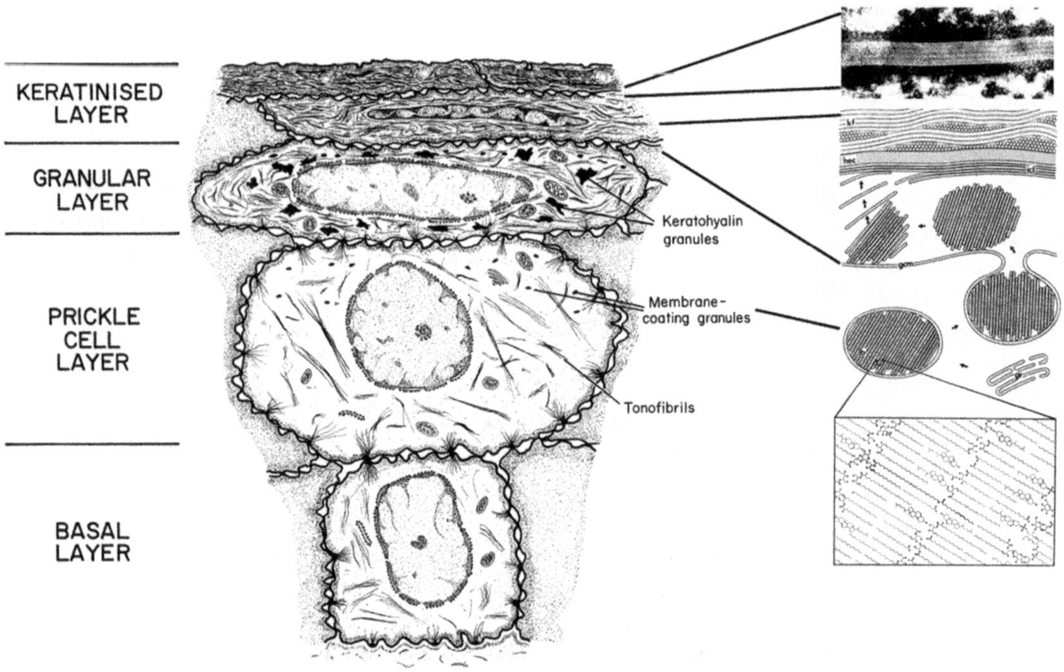

Fig. 1 Diagram to show the major structural changes during maturation and formation of the permeability barrier in keratinized epithelium. Insert: Membrane-coating granules are formed in the Golgi region (gr) in the prickle cell layer and migrate to the superficial cell membrane as cells reach the granular layer. Here, the membrane of the granule (gcm) fuses with the cell plasma membrane and the lipid lamellae (lb) are extruded into the intercellular space and become rearranged into intercellular lamellae (icl) that lie parallel to the horny cell envelope (hec) of the keratinized cell, containing keratin filaments (kf). We are grateful to Dr. Philip Wertz for providing the insert (reproduced from [5] with permission from Wiley)

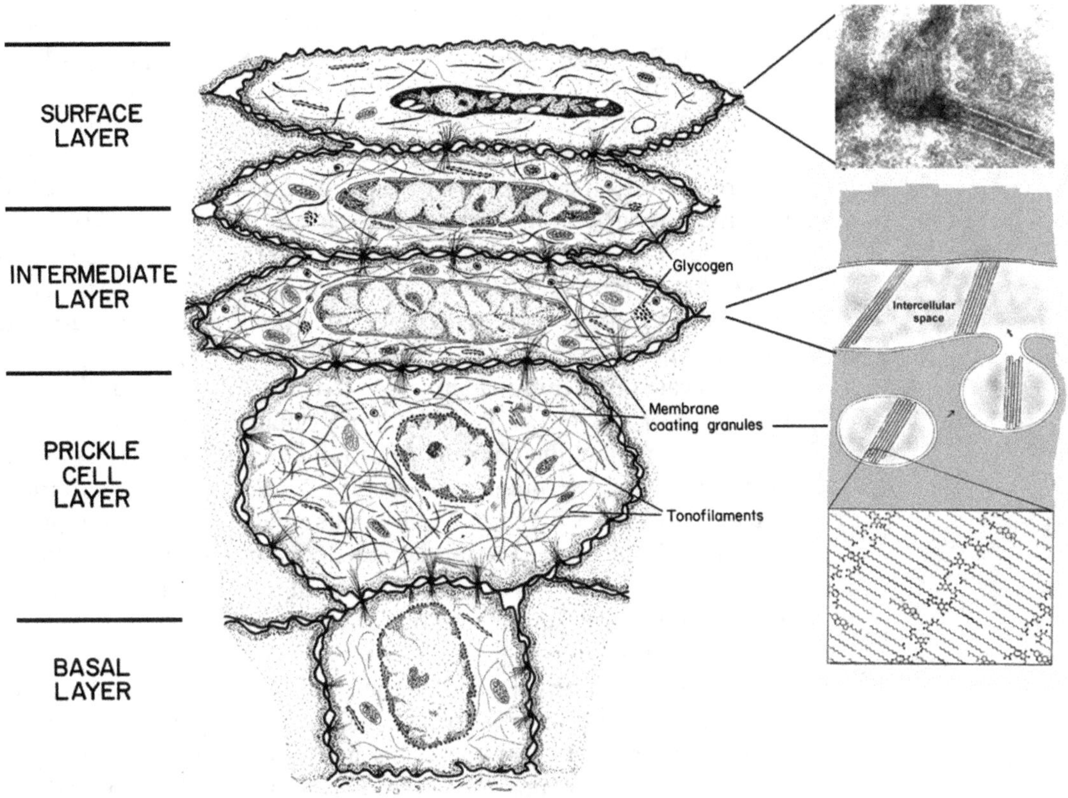

Fig. 2 Diagram to show the major structural changes during maturation and formation of the permeability barrier in non-keratinized epithelium. Membrane-coating granules containing short stacks of lipid lamellae and electron-lucent material migrate to the superficial cell membrane of cells in the intermediate layer of the epithelium. The membrane of the granule fuses with the cell plasma membrane and the lamellate contents are extruded into the intercellular space as shown in the electron micrograph in the top right panel. We are grateful to Dr. Philip Wertz for providing the panel insert (reproduced from [5] with permission from Wiley)

In non-keratinized epithelia such as that in the oral buccal and floor of mouth mucosa and the esophagus, vagina, and cervix (*see* Fig. 2), the accumulation of lipids is less evident than in keratinized epithelia. As cells reach the upper region of the epithelium, membrane-coating granules are again evident at the superficial aspect of the cells and appear to fuse with the plasma membrane so as to extrude their contents into the intercellular space. However, these granules are spherical in shape and membrane bounded [8] and have sometimes been referred to as "cored granules" because of their appearance in transmission electron micrographs. Only a small proportion of the granules in non-keratinized epithelium contain lamellae, which then appear scattered throughout the intercellular spaces in the outer portion of the epithelium [9]. It is the absence of extensive organized lipid lamellae in the intercellular spaces that may account for the greater permeability of non-keratinized tissues.

In order to study the penetration of antigens across skin or mucosa in the laboratory, investigators have utilized human, animal, and culture models. Among the latter, cell, tissue, and organ cultures have been used. Cultures of simple epithelia from the gastrointestinal tract, such as the Caco-2 cell line, have been proposed as a model for vaginal mucosa (reviewed in [10]) but it is important to remember that these do not reproduce the very different (and superior) barrier function of human vaginal stratified squamous epithelium. Organotypic culture models of skin, oral, and vaginal mucosa have been developed (reviewed in [10]) and show the structural characteristics of the prototypic human tissue but they do not usually reproduce the functional characteristics, such as a permeability barrier (reviewed in [10]).

Ethical considerations limit the study of pathogenic antigens in human subjects and it is difficult to obtain biopsies of normal human tissue in sufficient quantity for ex vivo studies. Thus attention has turned to animal models. The traditional small and inexpensive laboratory animals, such as mice, rats, and hamsters, have vaginal and oral mucosa that is keratinized and so does not resemble the human tissue. The oral and vaginal mucosa of rabbit has small areas that are non-keratinized but access is not easy. Larger laboratory animals such as dogs and monkeys do have oral and vaginal linings similar to human in structure and function, but these animals are expensive to procure and maintain. The pig has frequently been used for laboratory studies because of the striking similarity to humans in terms of anatomy, physiology, and metabolism, and this similarity extends to the oral and vaginal mucosae (reviewed in [10]). However, pigs are relatively expensive to maintain but their tissues are readily available from commercial slaughterhouses. We have shown that skin, oral, and vaginal mucosa can be successfully used for short-term studies of permeability ex vivo.

One method to measure the barrier function of a tissue is to determine the permeability of that tissue to compounds or organisms of interest using an in vitro perfusion system which includes temperature-controlled perfusion cells in which a tissue (or membrane) is mounted between a donor compartment (where the permeant is applied) and a receptor compartment which collects permeant that traverses the membrane. Perfusion systems can be static, which requires the manual sampling of receptor fluid, or a continuous flow system with automated sampling.

Static cells can be mounted vertically, as in the Franz-type cell (Fig. 3a), or side by side (Fig. 3b). The cells hold a fixed volume in

Fig. 3 Diagram of static cells showing (**a**) a vertical Franz cell and (**b**) a side-by-side cell. Both have a fixed volume stirred receptor compartment enclosed in a temperature-controlled water jacket and sampling port. The side-by-side cell allows stirring of both donor and receptor compartments (figure a and b courtesy of Andrew Wilt, PermeGear, Inc., Hellertown, PA). (**c**) Diagram of a continuous flow in-line cell showing a membrane clamped between the donor compartment and the receptor compartment through which receptor fluid is pumped (figure c courtesy of Andrew Wilt, PermeGear, Inc., Hellertown, PA)

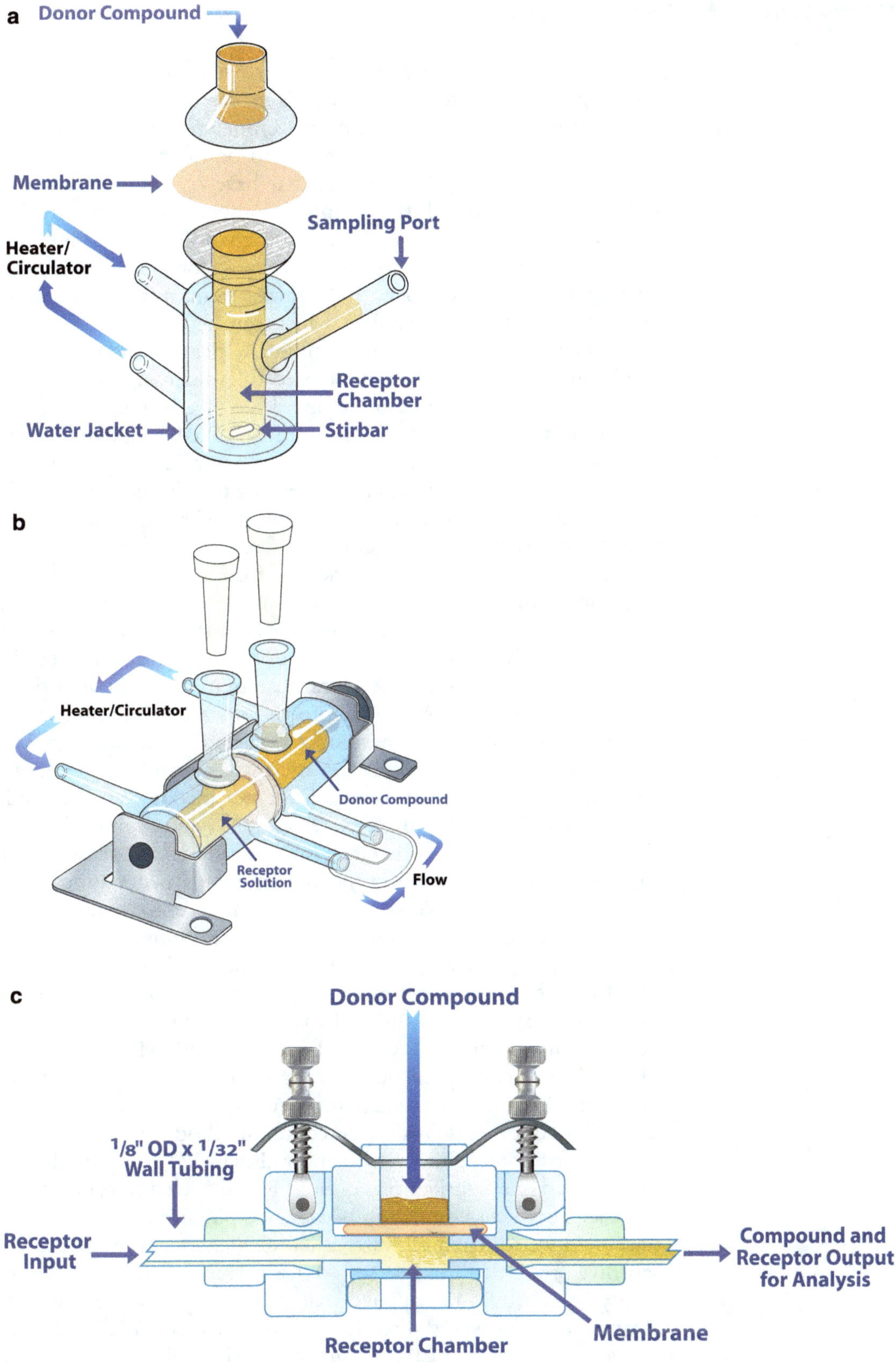
a
Donor Compound
Membrane
Sampling Port
Heater/
Circulator
Receptor
Chamber
Water Jacket
Stirbar
b
Heater/Circulator
Donor Compound
Receptor
Solution
Flow
c
Donor Compound
1/8" OD x 1/32"
Wall Tubing
Receptor
Input
Compound and
Receptor Output
for Analysis
Membrane
Receptor Chamber

the donor and receptor chambers and have a port through which to sample the receptor fluid. There is usually the capability to stir the receptor fluid and to control the temperature of the cell.

Continuous flow-thru cell scells can be of the vertical Franz-type cell or in-line cells (Fig. 3c) that have continuous flow of a collection fluid through the receptor chamber which causes turbulence and simulates stirring. The flow rate is adjustable, sampling is automatic, and temperature of the cells is controlled by placing them in thermostated blocks. Flow-thru systems minimize accumulation in the receptor chamber, which can be a concern for static cells where the permeant is not continually cleared, and reproduces the in vivo situation. They can be used to evaluate uptake into and across membranes at finite or infinite doses (see below).

Studies of tissue permeability can be valuable in determining whether, and under what conditions, a superantigen may penetrate a tissue. Alternatively, the penetration of a probe, such as water, in the presence of a superantigen, can indicate whether the tissue barrier is compromised. The agents of interest (e.g., superantigens or probes) can be prepared in a variety of formulations such as liquid solutions, suspensions, and gels, and applied as an infinite or as a finite dose. An infinite dose is one in which the permeant will not be significantly depleted from the donor formulation over the course of the experiment. A finite dose mimics the in vivo situation, where formulations applied to the surface of the mucosa do not usually persist for long periods of time and so become depleted.

The type of receptor solution used will depend on the nature and form of the permeant. An ideal receptor medium should mimic the in vivo situation. Lipid and water solubility and degree of ionization of a permeant must be taken into consideration so that it is in the desired state in the donor medium and, after penetration across the tissue, can be completely removed in the receptor medium.

The goal of a permeability study is to determine the amount of penetrant that is crossing the tissue. This can be described as a flux, which is the rate of diffusion of a permeant across the tissue or membrane expressed in units of mass/area/time. Alternatively, a permeability coefficient (K_p) can be determined, which is the rate of permeant penetration per unit concentration. The amount of permeant can be determined using unlabeled compounds with assays based on a sensitive detection method such as ELISA, HPLC, or fluorescence or by using radiolabeled compounds that are counted in a scintillation counter. Radiolabeled tracers provide high sensitivity and detection is usually rapid and accurate. As we have mentioned, tritiated water is an inexpensive, convenient, and sensitive probe that can be used to detect disruption of the permeability barrier by other applied materials. Penetration of bacterial toxins can be measured directly using internally radiolabeled toxin derived from microorganisms grown in vitro [11] or purchased commercially as custom-labeled compounds.

The continuous flow perfusion system permits investigation of environmental variables including bacterial virulence factors that may contribute to the penetration of a superantigen across skin and mucosa. It is possible to manipulate the tissue and environment in the perfusion chamber. The effect of elevated temperature (as might be experienced with fever), chemical damage from surfactants or other compounds found in topically applied products, mechanical damage that disrupts the epithelium, and therefore the permeability barrier, and the presence of toxins and other bacterial factors can be investigated in this system [12].

2 Materials

2.1 Equipment

The following account will describe the methodology for using radiolabeled toxin in an automated, continuous flow perfusion system (Fig. 4; described by Squier et al. [13] and manufactured by PermeGear, Inc., Hellertown, PA). The fraction collector (Fig. 4a) can be set to collect up to 18 samples with sampling intervals ranging from 1 min to several hours. It has a water-jacketed block (Fig. 4b) that can support 14 perfusion chambers and permits temperature control of the tissue specimens by means of a

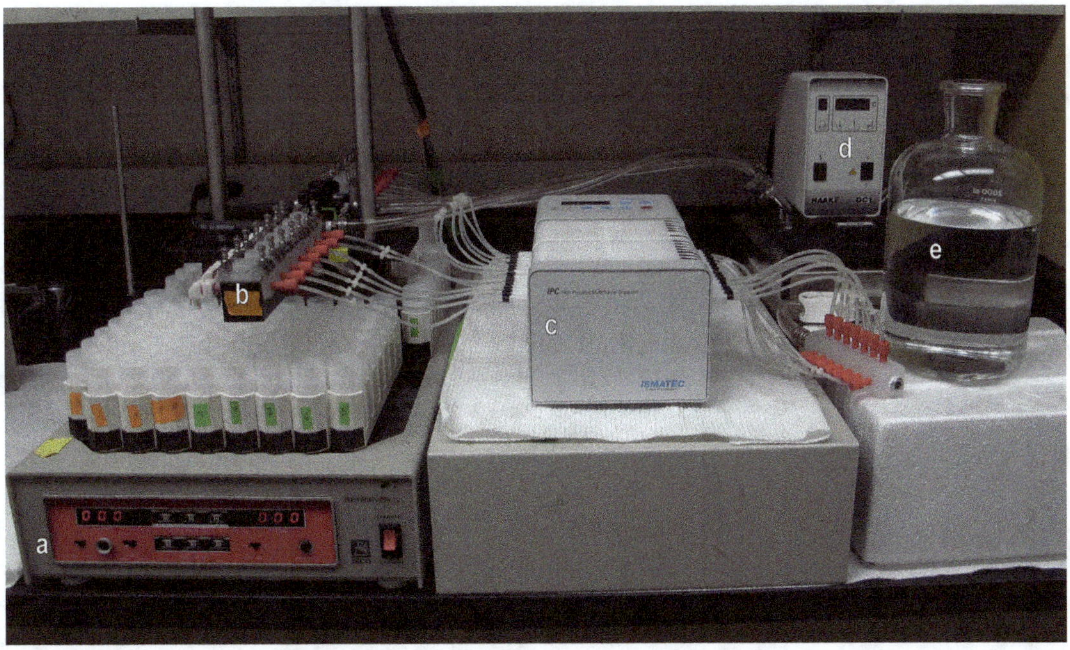

Fig. 4 Automated continuous flow perfusion system showing (*a*) fraction collector with collection vials in movable racks. Water-jacketed support blocks (*b*) holding perfusion cells are connected by tubing to the peristaltic pump (*c*) that delivers collection fluid from a reservoir (*e*). The temperature of the cells is controlled by the water pumped through the support blocks from the circulating water bath (*d*)

thermostated water bath (Fig. 4d). The peristaltic pump (Fig. 4c) is adjustable so as to obtain flow rates that are sufficient to clear perfusate from the receptor compartment of the chambers as it penetrates but provides a small enough volume to be adequately quenched by the scintillation fluid.

The perfusion chambers used in this system (Fig. 3c) accommodate small tissue biopsies of from 4 to 10 mm in diameter. The traditional Franz-type vertical chambers with a 9 mm diameter aperture used for skin penetration studies are too large for mucosa such as that from vagina, floor of mouth mucosa, and cheek, which tend to sag in the larger opening. Furthermore, some regions of oral mucosa from the porcine model, such as gingiva, floor of mouth, and palate, do not provide a sufficient tissue area for larger aperture chambers. Finally, small chambers provide the possibility of using human biopsy specimens obtained with a standard 4 mm diameter punch.

The donor compartment of the perfusion chamber has a capacity of 2 mL and has a large enough diameter to permit access to the exposed epithelial surface so as to be able to remove or change the applied materials. This is important if we wish to reproduce the in vivo situation where materials applied to the oral or vaginalmucosa do not remain for long periods of time because of tissue mobility and the flushing action of secretions.

2.2 Preparation of Solutions

1. Phosphate-buffered saline (PBS): 0.01 M, pH 7.4, 0.15 M NaCl. PBS is commonly used to deliver and collect permeants.
   (a) Prepare a 0.2 M monobasic stock solution: Dissolve 27.6 g of monobasic sodium phosphate, monohydrate, in distilled water and bring to final volume of 1 L.
   (b) Prepare a dibasic 0.2 M stock solution: Dissolve 28.4 g dibasic sodium phosphate in distilled water and bring to final volume of 1 L.
   (c) Prepare working solution: Mix 19.1 mL monobasic stock and 81.0 dibasic stock in about 1500 mL distilled water. Add 17.6 g NaCl, stirring until dissolved. Check pH and adjust if necessary to 7.4. Bring to a final volume of 2000 mL (*see* **Note 1**).
2. Donor solutions: Composition will vary by experiment.
   (a) Determine the nature of the donor solution (i.e., liquid buffer or gel medium) and the amount to apply per chamber. Liquid applications should be between 1 and 2 mL per chamber, and viscous applications between 100 and 500 μL per chamber (*see* **Note 2**).
   (b) Determine the amount of donor solution required for each condition. For example, prepare 8 mL for application of 1 mL liquid solution to seven chambers in an experi-

mental group plus enough extra to count multiple aliquots to determine the total cpm applied to each specimen as described in Subheading 3.5).

(c) Prepare control donor solution by adding the radiolabeled tracer to the selected donor medium to achieve a concentration of 1 μCi per application; such low concentrations are easily detectable in a scintillation counter but may not maintain an adequate diffusion gradient of the compound. In this case, the donor solution can be doped with unlabeled compound at a concentration consistent with the goals of study (*see* **Note 3**).

(d) Prepare the experimental donor solution(s) with the same components and concentrations as the control solution and then add the variable of interest, for example, a surfactant or a microbial virulence factor applied at concentrations that mimic the levels believed to be present in vivo.

(e) Determine the total activity of the donor solutions by adding 3 × 100 μL aliquots into scintillation fluid for counting (*see* **Note 4**).

2.3 Miscellaneous Supplies

1. Surgical retractor.
2. Scalpel, razor blades, and/or 10 mm biopsy punches.
3. Petri dishes.
4. Styrofoam sheet.
5. Dissecting pins.
6. Tweezers, scissor forceps.
7. Filter paper/parafilm.
8. Plastic 20 mL scintillation vials for perfusate collection.
9. Glass 20 mL scintillation vials for tissue solubilization.
10. Tissue solubilizer.
11. H_2O_2.
12. Scintillation cocktail.

2.4 Tissue Procurement

All tissue should be used as soon as possible after slaughter. Place the excised tissue in sealed plastic bags to keep the tissue moist and return to lab before trimming the tissue and cutting biopsies.

1. The vaginal tract is a tube that extends from the exterior vulva through the vagina, the uterine body, and uterine horns and ends at the ovaries. Remove the entire "tube" of tissue that includes the vagina and uterine body. Place the intact "tube" in plastic bag and return to lab before harvesting the vaginal mucosa.

2. Several oral mucosal tissues can be used in this system. Spread the pig jaw open with a surgical retractor. A fairly large area of buccal mucosa can be obtained from the inner cheek by cutting around the area with a razor blade and then, while holding an edge with blunt tweezers, slicing under the buccal mucosa to remove from the exterior cheek. A smaller area of floor of mouth mucosa (f/m), the loose, unattached mucosa extending from the underside of the tongue to the mandibular gingiva, is removed by cutting around the edges with a scalpel and then, while holding the edge, gently teasing away from underlying tissue with the scalpel tip. The attached gingiva is removed with a scalpel by running the scalpel tip under the largest gingival areas between incisor and premolar teeth.
3. Skin is removed from the glabrous region of the belly by cutting through the full thickness of skin with a razor blade and then excising it by slicing under the skin while pulling up on the edge with tweezers.

3 Methods

3.1 Preparation of Tissue

1. Vaginal mucosa: Open the vaginal tube with scalpel with one cut lengthwise. Pin the edges of the vaginal mucosa (the tissue closest to vulva) onto a piece of styrofoam and make one cut lengthwise to demarcate two sections (*see* **Note 5**). While holding the edge of the vaginal mucosa with blunt tweezers, excise the first mucosal sheet by carefully slicing under the mucosa as you gently pull up on it. Be careful not to stretch the tissue too drastically or cut into the epithelium as you tease the vaginal mucosa away from the underlying tissue (*see* **Note 6**). Put the vaginal mucosa sheets in a petri dish with buffer to keep it moist while removing the other mucosal sheet.
2. Oral mucosa: Cut strips of buccal mucosa about 1 cm wide, turn on the side, and trim away the submucosal tissue (*see* **Note 7**). Trim subcutaneous tissue from underside of f/m but leave intact until ready to cut biopsies. Gingiva does not need trimming. Keep all strips moist with buffer.
3. Skin: Cut strips of skin about 1 cm wide, turn on the side, and trim away the subcutaneous tissue (*see* **Note 8**).

3.2 Mounting Biopsies into Perfusion Chambers

1. Cut biopsies from the above-prepared tissue approximately 9 mm square with a sharp razor blade or 10 mm diameter with a punch (*see* **Note 9**).
2. Holding the tissue biopsy by the edges only with forceps and taking care not to puncture the epithelial surface, place it on the support button of the perfusion chamber and gently spread the biopsy to completely cover the opening (*see* **Notes 10** and **11**).

3. Place the donor compartment of the perfusion chamber on top of the biopsy, cover with the stainless steel strap, and secure with the threaded rods. Tighten the clamp securely with the locking nuts (*see* **Note 12**).
4. Place enough PBS in the donor compartment to cover the exposed epithelium and keep it moist, set aside, and continue to mount tissue in the number of perfusion chambers required for the experiment.

3.3 Setting Up the Experiment

1. Load the scintillation collection vials into the collection racks of the fraction collector and set the desired time intervals.
2. Turn on the circulating water bath and set the water temperature to heat the chambers to the desired experimental temperature of 37 °C.
3. Place the perfusion cells on the heated support bar of the fraction collector in replicates of seven per experimental condition.
4. Connect the intake tubing of the peristaltic pump to a reservoir of PBS (*see* **Note 13**).
5. Attach the other end of tubing from the peristaltic pump to the chambers and start pumping buffer though the receptor compartment.
6. Remove air bubbles from tubing and the receptor compartment of the chamber by tipping the chamber at a slight angle while pumping buffer at high speed until air bubbles are expelled (*see* **Note 14**).
7. Set the pump at a speed predetermined to remove the permeant at a sufficient rate to ensure sink conditions (*see* **Note 15**).
8. Remove the PBS from all donor compartments with vacuum or pipette carefully so as not to disturb the tissue which could damage the tissue or disrupt the seal.
9. At time zero, quickly but carefully pipette 1.0 mL of the prepared labeled solution to each donor compartment for both the control and experimental groups set up on the fraction collector.
10. Start the collection sequence set to the desired sample intervals, e.g. hourly.
11. Cover the donor compartment opening with a strip of parafilm to prevent evaporation.

3.4 Determine cpm in Perfusate Samples

1. Add 15 mL scintillation cocktail to the scintillation vials containing perfusate collected at intervals over the sampling period.
2. Count in a scintillation counter until a 2σ value of 2 % is obtained with automated quenching.

3.5 Determine cpm in Tissue and Donor Fluid (Mass Balance)

1. To determine the amount of labeled compound remaining in the donor compartment at the end of the experiment, collect the donor fluid and two rinses of the donor compartment into scintillation vials and count in scintillation counter as above.
2. To determine the amount of labeled compound retained in the tissue biopsies over the sampling period, remove the donor solution, rinse twice with buffer, and place the tissue into glass vials containing 2 mL of tissue solubilizer.
3. Gently agitate the tissue occasionally to aid digestion and keep the tissue from sticking to vial. The tissue can be mashed or cut into pieces to facilitate solubilization which can take several days at room temperature, or several hours incubated at 50 °C.
4. When all tissue is solubilized, add 100 μL of 30 % H_2O_2 to clear the slurry (may take multiple applications if highly colored).
5. Wait for 2 h with vials loosely capped, and when clear add 15 mL scintillation fluid and 150 μL glacial acetic acid (to eliminate chemiluminescence).
6. Wait a day before counting in a scintillation counter.

3.6 Preparation of Tissue for Microscopic Assessment

1. To assess histologic changes in the tissue over the course of the experiment you will need to prepare a control tissue (a piece of the tissue that was not placed in the perfusion chambers) and tissue specimens from each experimental group at the end of the experiment.
2. Carefully remove donor fluid and rinse with PBS so as not to damage the epithelial surface.
3. Remove the specimen from the chamber and blot dry with filter paper. You can use a biopsy punch to remove the flattened edges from the specimen leaving the exposed area of tissue.
4. The tissue biopsies can be prepared by placing them in embedding medium and frozen [14] for cutting thin sections in a cryostat [15] or by placing in 10 % formalin fixative for processing in wax [16].
5. Sections can then be stained with hematoxylin and eosin to assess changes in the tissue [17] (*see* **Note 16**).

3.7 Calculations and Statistical Analysis

1. Calculate the flux at each sample point for each chamber from the relationship flux = Q/At where Q is the quantity of compound traversing the tissue (cpm) in time t (min), and A is the area of epithelial surface exposed in cm^2. The units of flux are $cpm/cm^2/min$.
2. As the permeability coefficient by definition has to be calculated at steady state, this can be determined objectively by comparing flux values at sequential time intervals and determining the point at which there is no significant increase in

flux. These points represent accurate measurement of steady state. Calculate the mean flux at each sampling point within a group and compare consecutive flux values using ANOVA with an appropriate post test to identify values not significantly different at $p<0.05$, indicating a steady state. Calculate the mean steady-state values at steady state for each experimental group and compare using t-test for pairwise comparisons or ANOVA for multiple comparisons with an appropriate post test to identify values significantly different at $p<0.05$.

3. Calculate a permeability constant from the relationship $K_p = J/\Delta C$ where J is the flux at steady state and ΔC is the average concentration gradient expressed in cpm/cm^3. The units of K_p are cm/min. Statistical comparison is then made in the same way as with flux.

4 Notes

1. When running permeability with organisms, it is best to make fresh PBS rather than adding sodium azide for prevention of added bacterial growth.
2. We find 1 mL liquid applications to be satisfactory for ease of formulating without using excess materials; 100 μL is a sufficient amount of viscous application to completely cover the exposed epithelial surface which prevents drying of the tissue and keeps the compound of interest in close contact with the epithelial surface.
3. In our experience, 1 μCi applications ensure a sufficient diffusion gradient to detect low counts through tissue highly impermeable to compounds and ensure that enough counts remain in the donor solution throughout the experiment for highly permeable compounds.
4. Smaller aliquots of 20 or 50 μL would also work but counting and averaging triplicate aliquots gives a more accurate estimate of the cpm of the solution by reducing pipetting error.
5. From a histologic evaluation of tissue from different areas of the vaginal tract, the vaginal vestibule (the smooth mucosa near exterior vulva) structurally resembles that of human vaginal mucosa.
6. Do not be overly concerned with removing the underlying tissue at this point as you can remove it when you are preparing the individual biopsies in the next step.
7. It is important to prepare biopsies of consistent thickness. If available, a Padgett dermatome produces uniform buccal mucosa sheets.

8. The hair on the porcine belly skin should be carefully removed with a scissors, not scraped with a razor as this could remove the outer stratum corneum layer.
9. The tissue tends to "retract" or shrink once cut and we find that freehand cutting works better than a biopsy punch. You can cut a biopsy slightly oversize with a razor blade and then trim down as needed once the tissue retracts.
10. Size of tissue biopsies should neither be too large, which causes tissue to bulge into both halves of the perfusion chamber through the aperture, nor too small so that the tissue does not form a seal and will leak donor fluid into the receptor compartment.
11. While it is important to keep tissue moist prior to the start of the experiment, blotting the biopsy before putting into the perfusion chamber makes it easier to stretch and hold in place until it is clamped. Once it is clamped, apply buffer to the epithelial surface in the donor compartment.
12. The clamp should be tightened until snug but not overly tightened as it can cause the locking nuts to spring open.
13. Pump PBS through the tubing before attaching it to the perfusion chambers to remove air bubbles and determine that PBS is flowing through the tubing; no flow indicates that there are blockages or collapsed tubing and it needs to be replaced.
14. If the air bubbles will not exit the receptor chamber, this indicates that there is not a good seal around the tissue and it will leak.
15. The receptor compartment is quite small (<0.3 mL) in relation to the donor compartment (2 mL) and we have found that, for most compounds, pumping buffer at 1.5 mL per hour is sufficient to remove the compound penetrating the mucosal tissue from the receptor compartment and also maintain a concentration gradient.
16. If the tissue has been exposed to radiolabeled compounds, the tissue needs to be processed by people trained to handle radioactivity and the processing and sectioning equipment must be designated for radioactive use.

References

1. Smith AJ, Jackson MS, Bagg J (2001) The ecology of *Staphylococcus* species in the oral cavity. Rev J Med Microbiol 50:940–946
2. Human Microbiome Project Consortium (2012) Structure, function and diversity of the healthy human microbiome. Nature 486:207–214. doi:10.1038/nature11234
3. Guinan ME, Dan BB, Guidotti RJ, Reingold AL, Schmid GP, Bettoli EJ, Lossick JG, Shands KN, Kramer MA, Hargrett NT, Anderson RL, Broome CV (1982) Vaginal colonization with Staphylococcus aureus in healthy women: a review of four studies. Ann Intern Med 96:944–947
4. Norris D (1989) Immunological cytotoxicity of cutaneous targets. In: Norris DA (ed) Immune mechanisms in cutaneous disease. Marcel Dekker, New York, pp 123–166

5. Squier C, Brogden KA (eds) (2011) Human oral mucosa: development, structure, & function. Wiley, New Jersey, NJ
6. Martinez IR, Peters A (1971) Membrane-coating granules and membrane modifications in keratinizing epithelia. Am J Anat 130:93
7. Landmann L (1988) The epidermal permeability barrier. Anat Embryol 178:1–13
8. Squier CA (1977) Membrane coating granules in nonkeratinized oral epithelium. J Ultrastruct Res 60:212–220
9. Wertz PW, Swartzendruber DC, Squier CA (1993) Regional variation in the structure and permeability of oral mucosa and skin. Adv Drug Del Rev 12:1–12
10. Squier CA, Mantz MJ, Schlievert PM, Davis CC (2008) Porcine vagina *ex vivo* as a model for studying permeability and pathogenesis in mucosa. J Pharm Sci 97:9–21
11. Davis CC, Kremer MJ, Schlievert PM, Squier CA (2003) Penetration of toxic shock syndrome toxin-1 across porcine vaginal mucosa ex vivo: permeability characteristics, toxin distribution and tissue damage. Am J Obstet Gynecol 189:1785–1789. doi:10.1016/S0002-9378(03)00873-1
12. Davis CC, Baccam M, Mantz MJ, Osborn TW, Hill DR, Squier CA (2014) Use of porcine vaginal tissue ex-vivo to model environmental effects on vaginal mucosa to toxic shock syndrome toxin-1. Toxicol Appl Pharmacol 274:240–248. doi:10.1016/j.taap.2013.11.021
13. Squier CA, Kremer M, Wertz PW (1997) Continuous flow mucosal cells for measuring the in-vitro permeability of small tissue samples. J Pharm Sci 86:82–84
14. Peters S (ed) (2010) Frozen block cryoembedding In: A practical Guide to Frozen Section Technique. Springer Science. http://pathologyinnovations.com/frozen_section_technique.htm. Accessed 29 Sept 2014
15. Peters S (ed) (2010) Frozen section technique. In: A practical Guide to Frozen Section Technique. Springer Science. http://pathologyinnovations.com/frozen_block_embedding.htm. Accessed 29 Sept 2014
16. Paraffin processing of tissue (2012) In: Protocols Online. http://protocolsonline.com/histology/sample-preparation/paraffin-processing-of-tissue/. Accessed 29 Sept 2014
17. Ellis R (2011) Hematoxylin and Eosin (H&E) Staining Protocols. IHC World. http://www.ihcworld.com/_protocols/special_stains/h&e_ellis.htm. Accessed September 29, 2014

Chapter 13

Protocol for Examining Human Vaginal Epithelial Cell Signaling in Response to Staphylococcal Superantigens

Laura M. Breshears and Marnie L. Peterson

Abstract

A detailed investigation of eukaryotic signaling pathways affected by bacterial products is key to our understanding of host-pathogen interactions. Cytokine expression appears to be an important initial host cell response to many bacterial products, including the *Staphylococcus aureus* superantigens (SAgs). While much is understood about how SAgs signal to immune cells, very little is known about the specific cellular pathways activated by SAgs on nonimmune cells such as those of the epithelium. Here, we describe methods for analyzing SAg signaling in cultured epithelial cells, which may be extrapolated to the analysis of signaling pathways induced by other bacterial ligands on a variety of cell types.

Key words Cell signaling, Ectodomain shedding, Superantigen, *Staphylococcus aureus*, Toxin, Epithelium, Mucosa

1 Introduction

Signaling pathways induced in host cells by bacterial ligands are often the initial steps in host-pathogen interactions. Describing how bacterial ligands interact with host cell receptors and induce intracellular signaling cascades can provide drug targets that take advantage of our knowledge of host biology in attempting to reduce infections by virulent organisms. Staphylococcal superantigens (SAgs) cause disease through hyper-activation of the immune system leading to massive cytokine release and systemic shock [1]. The SAg toxic shock syndrome toxin-1 (TSST-1) causes menstrual toxic shock syndrome (mTSS) [2, 3]. TSST-1 binds to epithelial cells, is quickly internalized, and disrupts the integrity of the mucosal epithelium [2, 4–6]. In a cell culture model, TSST-1 causes cell rounding and the production of pro-inflammatory cytokines and chemokines from human vaginal epithelial cells (HVECs) [5]. Evidence suggests that local inflammation induced by cytokine

Amanda J. Brosnahan (ed.), *Superantigens: Methods and Protocols*, Methods in Molecular Biology, vol. 1396,
DOI 10.1007/978-1-4939-3344-0_13, © Springer Science+Business Media New York 2016

release from mucosal epithelial cells is necessary for disease progression in mTSS [7].

The mechanism of cytokine release via interactions between TSST-1 and epithelial surfaces was recently described [8]. Analyses of molecules shed from the HVEC surface and secreted from cells into cell culture medium in response to TSST-1 were performed via enzyme-linked immunosorbent assay (ELISA). Western blotting of whole-cell lysates was used to determine the activation state of the intracellular signaling molecule ERK1/2 by investigating its phosphorylation state. Various commercially available small-molecule inhibitors (AG1478 and TAPI-1) were used to confirm the involvement of transmembrane signaling molecules such as the epidermal growth factor receptor (EGFR), and a disintegrin and metalloproteinases (ADAMs), respectively. Activation of intracellular signaling molecules such as Gβγ and MEK1/2 was also confirmed with inhibitors (Gallein and U0126, respectively). All of these data allowed for a detailed description of the signaling cascade induced on HVECs by TSST-1 and provided novel drug targets for the prevention of SAg-mediated diseases.

2 Materials

2.1 Cell Culture and Lysis

1. HVECs obtained from ATCC (CRL-2616) [9].
2. Complete medium: Keratinocyte serum-free medium (KSFM) supplemented with 0.2 ng/ml human recombinant EGF, 50 μg/ml bovine pituitary extract (GIBCO, Grand Island, NY), 0.4 mM $CaCl_2$, 25 IU/ml penicillin, 25 μg/ml streptomycin, 40 μg/ml gentamicin, and 2.5 μg/ml amphotericin B.
3. Minimal medium: KSFM supplemented only with 0.4 mM $CaCl_2$.
4. Trypsin-EDTA.
5. Sterile molecular biology-grade PBS supplemented with 10 % fetal bovine serum (FBS).
6. RIPA lysis buffer: 150 mM NaCl, 100 mM Tris–HCl [pH 7.5], 0.5 % deoxycholate, 0.1 % sodium dodecyl sulfate, 1 % Triton X-100, store at 4 °C (*see* **Note 1**).
7. Protease inhibitor (i.e., Roche complete, EDTA-free protease inhibitor tablets) (*see* **Note 2**).
8. Tissue culture-treated flasks, 6- and 96-well plates.
9. Sterile non-tissue culture-treated 96-well plates.
10. Biosafety cabinet for manipulating cells in a sterile environment.
11. Incubator for growth at 37 °C + 7 % CO_2.

2.2 ELISA Reagents

1. ELISA kits (for example, obtained from R&D Systems, Inc., Minneapolis, MN).
2. Wash buffer: PBS + 0.05 % Tween-20 (*see* **Note 3**).
3. 96-Well flat-bottom ELISA plates.
4. ELISA plate reader.

2.3 Western Blotting

1. Mini PROTEAN Tetra cell electrophoresis chamber (Bio-Rad or comparable brand).
2. Mini Trans-blot cell (Bio-Rad or comparable brand).
3. Precast 4–20 % 10-well gradient SDS-PAGE gels.
4. Laemmli protein sample buffer + β-mercaptoethanol.
5. Pre-stained ladder.
6. PVDF membrane.
7. Extra thick blot paper.
8. SDS-PAGE running buffer: 25 mM Tris base, 192 mM glycine, 0.1 % SDS, pH 8.0 (*see* **Note 4**).
9. Western blot transfer buffer: 25 mM Tris base, 192 mM glycine, 20 % MeOH (*see* **Note 5**).
10. Wash buffer: Tris-buffered saline + Tween (TBST), 50 mM Tris base, 150 mM NaCl, pH 7.4 + 0.05 % Tween-20 (*see* **Note 6**).
11. Block buffer: TBST + 5 % milk + normal goat serum, store at 4 °C (*see* **Note 7**).
12. Antibody diluent: TBST + 5 % milk with 0.02 % sodium azide, store at 4 °C (*see* **Note 8**).
13. Western stripping buffer: 200 mM glycine, 0.1 % SDS, 1 % Tween-20, pH 2.2 (*see* **Note 9**).
14. Primary antibodies: Mouse anti-ERK1/2 and rabbit anti-phospho-ERK1/2 (for example, from Santa Cruz Biotechnology).
15. Horse anti-mouse HRP-linked detection antibody (*see* **Note 10**).
16. Goat anti-rabbit HRP-linked detection antibody (*see* **Note 10**).
17. SuperSignal West Pico Chemiluminescent Substrate kit (Pierce or comparable brand).
18. X-ray film.

2.4 Inhibitors and Stimulants (See Notes 11 and 12)

1. AG1478 (Tocris): 1 mM stock in DMSO.
2. U0126 (Tocris): 10 mM stock in DMSO.
3. Gallein (Tocris): 10 mM stock in DMSO.

4. TAPI-1 (Enzo Life Sciences): 5 mM stock in DMSO.
5. TSST-1: 1 mg/ml stock in sterile H_2O.
6. TGFα: 100 μg/ml stock in 10 mM acetic acid.

3 Methods

3.1 Analysis of Ectodomain Shedding and Cytokine Secretion via ELISA

3.1.1 HVEC Preparation (All Steps Performed in Biosafety Cabinet)

1. Trypsinize a T225 or T75 flask depending on how many cells are required. A T75 that is ~80 % confluent will have ~3×10^6 total cells. A T225 that is ~80 % confluent will have ~1×10^7 total cells (*see* **Note 13**).
2. Dilute cells to 5×10^5 cells/ml in complete KSFM.
3. For each ELISA to be done and stimulation condition to be tested, *see*d 100 μl/well (this is 50,000 cells/well) in triplicate into a 96-well tissue culture-treated plate (*see* **Note 14**).
4. After 24 h, replace the complete medium covering the cells with 100 μl/well minimal medium.
5. After 24 h the cells are ready to be treated and processed.

3.1.2 Preparation of Reagents, HVEC Stimulation, and ELISAs

1. Thaw inhibitors and TSST-1 on the bench top or in a 37 °C water bath, vortex briefly to mix, and centrifuge briefly to collect contents in the bottom of tubes.
2. In a biosafety cabinet, dilute inhibitors or TSST-1 in appropriate volume of minimal medium such that you have 50 μl/well of 2× concentrated inhibitor or 2× concentrated TSST-1 (200 μg/ml) for each well that will receive toxin or a particular inhibitor.
3. Aspirate medium from cells, add 50 μl of inhibitors or minimal medium to appropriate wells, and incubate cells for 30 min (*see* **Note 15**).
4. Stimulate HVECs by adding 50 μl TSST-1 (or minimal medium for medium-alone control wells) and incubate for 6 h.
5. Using a multichannel pipette, remove supernates from each well and transfer to a sterile non-tissue culture-treated 96-well plate. Wrap plate in parafilm and freeze supernates at −20 °C until the day of the ELISA.
6. ELISAs are always performed exactly according to the manufacturer's instructions. ELISA data is usually given in pg/ml or ng/ml. Depending on the molecule under investigation, there may be no detectable protein in the cell supernates in the medium-alone controls and there can be wide day-to-day variations in the TSST-1-stimulated wells. This precludes normalization and combination of data from multiple days; therefore representative data should always be shown.

3.2 Analysis of ERK1/2 Phosphorylation

3.2.1 HVEC Preparation (All Steps Performed in Biosafety Cabinet)

1. Trypsinize cells as described above and dilute to 5×10^5 cells/ml in complete KSFM.
2. For each time point and stimulation condition to be tested, add 200 μl/well (100,000 cells) to each well of a 24-well tissue culture plate (*see* **Note 16**).
3. After 24 h, replace the complete medium covering the cells with 200 μl/well minimal medium.
4. After 24 h the cells are ready to be treated and processed.

3.2.2 Preparation of Reagents, HVEC Stimulation, and Cell Lysis

1. Thaw inhibitors, TSST-1, and TGFα, on the bench top or in a 37 °C water bath, vortex briefly to mix, and centrifuge briefly to collect contents in the bottom of tubes. All subsequent steps should be performed in a biosafety cabinet.
2. Dilute inhibitors or TSST-1 in appropriate volume of minimal medium such that you have 150 μl/well of 2× concentrated inhibitor or 2× concentrated TSST-1 (200 μg/ml) for each well that will receive toxin or a particular inhibitor.
3. Dilute TGFα to 10 μg/ml in minimal medium.
4. Aspirate media from cells, add 150 μl of inhibitors or minimal medium to appropriate wells, and incubate cells for 30 min (*see* **Note 15**).
5. Stimulate HVECs by adding 150 μl TSST-1 and incubate for 5–30 min.
6. TGFα acts as a positive control for ERK1/2 phosphorylation. Add 300 μl to one well and incubate for 5 min.
7. Quickly aspirate media from cells at each time point and add 150 μl RIPA lysis buffer with protease inhibitors.
8. Incubate for 2 min and transfer lysate to microcentrifuge tubes, pipetting up and down in well to break up clumps and ensure maximum removal of protein.
9. Centrifuge at maximum speed in a microcentrifuge for 2 min to pellet cell debris.
10. Transfer supernates to fresh microcentrifuge tubes and store at –20 °C.

3.2.3 Western Blotting forPhosphoERK1/2

1. We tested TSST-1 alone, and TSST-1 + AG1478 *or* TAPI-1; therefore the following protocol is for three western blots.
2. Thaw cell lysates from previous section on ice.
3. Remove 50 μl of each lysate from a TSST-1-stimulated sample and mix with 50 μl protein sample buffer.
4. Remove 100 μl of lysate from the TGFα-stimulated sample and mix with 100 μl protein sample buffer.
5. Return remaining lysates to –20 °C for future use if necessary.

6. Boil samples in microcentrifuge tubes in a heating block at 100 °C for 5 min.
7. Centrifuge all samples at maximum speed in a microcentrifuge for ~30 s.
8. Prepare three precast protein gels by removing the strips of tape below the gel and the comb from the top of the gel. Gently rinse with diH_2O to remove air bubbles from the wells.
9. Assemble the SDS-PAGE gel-running chamber and fill with running buffer.
10. Add 10 μl of pre-stained ladder to the first well on the left of each gel.
11. Moving left to right, add 25 μl of each cell lysate sample beginning with the 0 time point. One gel will contain samples from TSST-1 alone, the second gel will contain samples from TSST-1 + AG1478, and the third gel will contain the TSST-1 + TAPI-1 samples. The TGFα sample should be included on each gel as a positive control.
12. Run the gel at 100 V for ~ 1 h or until the dye front reaches the bottom of the gel.
13. Prepare for transfer by presoaking transfer paper and pads in transfer buffer. Wet PVDF membranes in MeOH and rinse vigorously with transfer buffer until the membranes sink in the buffer.
14. If using the BioRad apparatus, assembly is as follows (if not, follow your manufacturer's instructions): place a transfer cassette with the black side down on the bench. Place a wet pad on the black side of the cassette followed by a wet piece of transfer paper. When the gel is finished, pry the plastic gel plates apart with the metal key provided. Remove the wells with a razor blade and carefully transfer the gel to the sandwich assembly, placing it on top of the transfer paper with the ladder facing up and on the right (the ladder will then transfer to the left side of the membrane allowing for left-to-right sample visualization). Take care not to allow the gel to dry out. Place a membrane over the gel and use a clean glass pipette to gently roll any bubbles from between the membrane and the gel. Finish assembly by placing another piece of transfer paper over the membrane (again, rolling a glass pipette across to squeeze out any bubbles) and another wet pad over the paper. Close the cassettes and place into the transfer box with an ice block and a small stir bar.
15. Transfer the proteins to the membranes at 100 V for 1 h.
16. Wash membranes with TBST on rocker for 10 min.
17. Block membranes for 1 h at room temp.

18. Probe membranes with mouse α–ERK1/2 primary antibody at 1:1000 in antibody diluent overnight at 4 °C with gentle rocking.
19. Remove and save primary antibody if it contains sodium azide.
20. Wash blots 2× quickly and then 3 × 5 min with TBST.
21. Incubate membranes with horse α–mouse HRP secondary antibody at 1:4000 in TBST for 1 h at room temp with gentle rocking.
22. Wash membranes 2× quickly and then 3 × 5 min with TBST.
23. Place membrane on plastic wrap, add 1 ml of mixed substrate, and incubate for 1 min.
24. Transfer membranes to darkroom cassette and expose X-ray film for the appropriate amount of time (usually 30 s–10 min) until bands are visible and results clear.
25. Strip antibodies from membranes: wash 2 × 5–10 min with stripping buffer followed by 2 × 10 min PBS and then 2 × 5 min TBST.
26. Repeat **steps 17–26** above with rabbit α–phospho-ERK1/2 primary antibody at 1:1000, and goat α–rabbit HRP secondary antibody at 1:30,000.

4 Notes

1. RIPA lysis buffer can be made ahead of time and stored at 4 °C for up to 1 month.
2. Protease inhibitors should be made fresh, kept on ice, and added to lysis buffer just prior to cell lysis for every experiment. Care should be taken to store protease inhibitors properly and discard at the expiration date.
3. ELISA wash buffer does not need to be sterile if made often.
4. A 10× solution of running buffer can be used to quickly make 1 L of 1×. The 10× solution is made without SDS. Add 100 ml 10× buffer to 890 ml dH_2O, and then add 10 ml 10 % SDS. Running buffer can be reused several times.
5. A 10× solution of transfer buffer can be used to quickly make 1 L of 1×. The 10× solution is made without methanol. Add 100 ml 10× buffer to 700 ml dH_2O, and then add 200 ml methanol. Running buffer can be reused several times and should be stored at 4 °C in the dark.
6. A 10× solution of wash buffer can be used to quickly make 1 L of 1×. The 10× solution is made without Tween-20. Add 100 ml 10× buffer to 900 ml dH_2O and then add 0.5 ml Tween-20 while stirring on a stir plate.

7. Blocking buffer can be used over the course of ~3 days if stored at 4 °C. The shelf life can be extended by the addition of sodium azide (0.02 %) to keep contaminants from growing as long as it will not interfere with downstream steps. A quick smell will inform the user if the buffer has grown contaminants and should be made fresh.
8. Antibodies can be diluted in this diluent and reused multiple times. Some polyclonal antibodies will show greater specificity over multiple uses as nonspecific reactions are diminished. The addition of sodium azide is critical to block growth of contaminants. If sodium azide is omitted the antibody should be made fresh every time. Sodium azide is highly toxic and should be used according to its MSDS safety guidelines.
9. Many western blots can be stripped of antibodies and re-probed up to ten times. Stripping requires a low pH. Make 1 L of stripping buffer by mixing 15 g glycine with 1 g SDS and 10 ml Tween-20 into 800 ml diH_2O. Add 10 M HCL until pH 2.2 is reached (~10 ml) and then bring solution to 1 L with diH_2O.
10. Sodium azide is a potent inhibitor of HRP. Membranes should be washed thoroughly in wash buffer to remove residual azide from the blocking/probing steps. The HRP detection antibody is diluted in wash buffer just prior to use.
11. All inhibitor stocks were made with sterile molecular biology-grade DMSO aliquoted in a biosafety cabinet. The stock concentrations of inhibitors are 1000× the working concentrations for cell culture experiments. Final working concentration for each inhibitor is AG1478 = 1 μM, U0126 = 10 μM, Gallein = 10 μM, TAPI-1 = 50 μM. The final working concentration of TSST-1 on HVECs is 100 μg/ml. The final working concentration of TGFα on HVECs is 10 μl.
12. Inhibitors should be stored in 1 ml aliquots at −20 °C. Freeze-thawing inhibitors multiple times had no affect on activity.
13. All cell growth and incubations must be performed in a 37 °C incubator with 7 % CO_2. Trypsinize by aspirating media from flasks, washing 1 × 5 ml (T75) or 10 ml (T225) with sterile PBS, and incubating for ~10 min with 5 ml (T75) or 10 ml (T225) trypsin-EDTA. Dislodge cells completely by tapping the side of the flasks against a flat palm. Remove cells to 10 or 20 ml tubes. Wash flasks with 5 ml (T75) or 10 ml (T225) PBS + 10 % FBS and add wash to cells in tubes. Cap tightly and spin to pellet cells. Aspirate off medium and resuspend cells by vortexing in 5 ml (T75) or 10 ml (T225) fresh complete KSFM. Count cells (we use a hemocytometer) and use this number for determining the dilution factor necessary for next steps.

14. For example, if you wish to test secretion of IL-8 and shedding of two surface molecules in the presence or absence of SAgs with and without two inhibitors, then you will be doing three ELISAs to test four conditions: media alone, SAg alone, SAg + inhibitor 1, and SAg + inhibitor 2. Each condition requires an $n=3$, so 3 ELISAs × 4 conditions × 3 replicates = 36 HVEC-seeded wells to complete the experiment.
15. Adding inhibitors and toxins in ½ the final volume at 2× the final concentration greatly simplifies pipetting. TSST-1-alone wells receive 50 μl/ well medium at the time of addition of inhibitors. Adding 2× concentrated TSST-1 to wells with inhibitors dilutes both the inhibitor and the toxin to the appropriate final dilution. Adding 2× concentrated TSST-1 to wells with 50 μl medium dilutes the toxin to the appropriate final dilution. Be sure to always include medium-alone control wells to show basal, un-stimulated activity.
16. For example, we tested TSST-1 +/− AG1478 or TAPI-1 at 0, 5, 10, 15, and 30 min with a 5-min TGFα stimulation as a positive control. This required 16 wells total: 5 for TSST-1 alone, 5 for TSST-1 + AG1478, 5 for TSST-1 + TAPI-1, and 1 for TGFα.

Acknowledgements

This work was supported, in whole or in part, by National Institutes of Health Grants R01AI073366-03 and R01AI074283 through the NIAID. We would like to thank Dr. Matthew Schaefers for helpful discussions and technical assistance in executing this work. Dr. Patrick Schlievert, University of Iowa, provided the TSST-1 protein.

References

1. McCormick JK, Yarwood JM, Schlievert PM (2001) Toxic shock syndrome and bacterial superantigens: an update. Annu Rev Microbiol 55:77–104
2. Davis CC, Kremer MJ, Schlievert PM, Squier CA (2003) Penetration of toxic shock syndrome toxin-1 across porcine vaginal mucosa ex vivo: permeability characteristics, toxin distribution, and tissue damage. Am J Obstet Gynecol 189:1785–1791
3. Schlievert PM, Shands KN, Dan BB, Schmid GP, Nishimura RD (1981) Identification and characterization of an exotoxin from Staphylococcus aureus associated with toxic-shock syndrome. J Infect Dis 143:509–516
4. Kushnaryov VM, MacDonald HS, Reiser R, Bergdoll MS (1984) Staphylococcal toxic shock toxin specifically binds to cultured human epithelial cells and is rapidly internalized. Infect Immun 45:566–571
5. Peterson ML, Ault K, Kremer MJ, Klingelhutz AJ, Davis CC, Squier CA, Schlievert PM (2005) The innate immune system is activated by stimulation of vaginal epithelial cells with Staphylococcus aureus and toxic shock syndrome toxin 1. Infect Immun 73:2164–2174

6. Shupp JW, Jett M, Pontzer CH (2002) Identification of a transcytosis epitope on staphylococcal enterotoxins. Infect Immun 70:2178–2186
7. Brosnahan AJ, Schaefers MM, Amundson WH, Mantz MJ, Squier CA, Peterson ML, Schlievert PM (2008) Novel Toxic Shock Syndrome Toxin-1 Amino Acids Required for Biological Activity. Biochemistry 47:12995–13003
8. Breshears LM, Schlievert PM, Peterson ML (2012) A Disintegrin and Metalloproteinase 17 (ADAM17) and Epidermal Growth Factor Receptor (EGFR) Signaling Drive the Epithelial Response to Staphylococcus aureus Toxic Shock Syndrome Toxin-1 (TSST-1). J Biol Chem 287:32578–32587
9. Fichorova RN, Rheinwald JG, Anderson DJ (1997) Generation of papillomavirus-immortalized cell lines from normal human ecto-cervical, endocervical, and vaginal epithelium that maintain expression of tissue-specific differentiation proteins. Biol Reprod 57:847–855

Chapter 14

Organ Culture as a Model System for Studies on Enterotoxin Interactions with the Intestinal Epithelium

Ulver Spangsberg Lorenzen, Gert H. Hansen, and E. Michael Danielsen

Abstract

Studies on bacterial enterotoxin-epithelium interactions require model systems capable of mimicking the events occurring at the molecular and cellular levels during intoxication. In this chapter, we describe organ culture as an often neglected alternative to whole-animal experiments or enterocyte-like cell lines. Like cell culture, organ culture is versatile and suitable for studying rapidly occurring events, such as enterotoxin binding and uptake. In addition, it is advantageous in offering an epithelium with more authentic permeability/barrier properties than any cell line, as well as a subepithelial lamina propria, harboring the immune cells of the gut mucosa.

Key words Organ culture, Enterotoxins, Brush border, Enterocytes, Small intestine, Mucosal explants

1 Introduction

Enterotoxins are proteins produced by invasive pathogenic microorganisms and secreted in the intestinal lumen to weaken the defense of the host organism with the aim to facilitate colonization. Infections with enterotoxin-producing bacteria typically cause secretory diarrhea, vomiting, and in severe cases toxic shock and death, and collectively they are responsible for a huge worldwide morbidity and mortality [1]. Whereas some enterotoxins, for instance those of *V. cholerae* and enterotoxigenic *E. coli*, directly target the absorptive epithelium [2], others, like those of *S. aureus*, primarily target the immune cells in the underlying lamina propria by acting as "superantigens" [3]. By tradition, studies on enterotoxin-epithelium interactions most often rely either on animal experiments or epithelial cell cultures, but in this chapter we describe organ culture as an alternative model system to be considered. Introduced several decades ago for human biopsies by Browning and Trier [4], organ culture may keep intestinal tissue viable for periods up to 24 h, permitting studies on metabolism,

Amanda J. Brosnahan (ed.), *Superantigens: Methods and Protocols*, Methods in Molecular Biology, vol. 1396,
DOI 10.1007/978-1-4939-3344-0_14, © Springer Science+Business Media New York 2016

cell proliferation, and nutrient absorption. Applying the technique to porcine mucosal explants [5], we have used organ culture for a number of years in studies of a great many processes related to various aspects of gut mucosal biology. Feeling that organ culture today is being somewhat neglected by experimental researchers in the field of gastroenterology, and based on our own experience, we wish to present this technique as an attractive candidate model system for studies on enterotoxin-epithelium interactions.

Like cell culture, organ culture is versatile and is suitable for short-term exposures (on the minute scale), when early events in the enterotoxin-epithelium interaction are under investigation. In addition, by comparison with cell culture, organ culture offers two more advantages:

1. A native epithelium displaying authentic permeability/barrier properties: Though polarized, enterocyte-like cell lines (for instance Caco-2 and HT-29) do not possess an apical, fully developed brush border including a subapical, filamentous terminal web and a full complement of digestive enzymes and nutrient transporters at the cell surface.
2. An in vivo-like epithelial organization possessing an underlying lamina propria containing cells of the immune system in the gut: This makes it feasible to study for instance synthesis, epithelial transcytosis, and apical secretion of locally produced immunoglobulins [6, 7].

We have used organ culture of porcine jejunal mucosal explants as a model system to study binding and uptake of enterotoxins from *V. cholerae* and *S. aureus* [8, 9]. In these works, we were able to show that cholera toxin subunit B binds to lipid raft microdomains at the enterocyte brush border and is taken up by an induced clathrin-dependent mechanism (Fig. 1) [9]. In contrast, the two staphylococcal enterotoxins, SEA and SAB, only associated poorly with lipid rafts but were taken up into a distinct population of early endosomes in the terminal web region (TWEEs), from where they engaged in transcytosis to reach their target cells in the subepithelial lamina propria (Fig. 2) [8].

Fig. 1 (continued) whereas in others, both enterotoxins (*yellow arrows*) were taken up into the same endosomes. The labeling pattern suggests that although CTB and HLT bind to a common receptor, ganglioside GM_1, at the cell surface, they exhibit some variation in their interaction profile with the intestinal epithelium. Nuclei were visualized by DAPI in the merged image. Bar: 10 μm

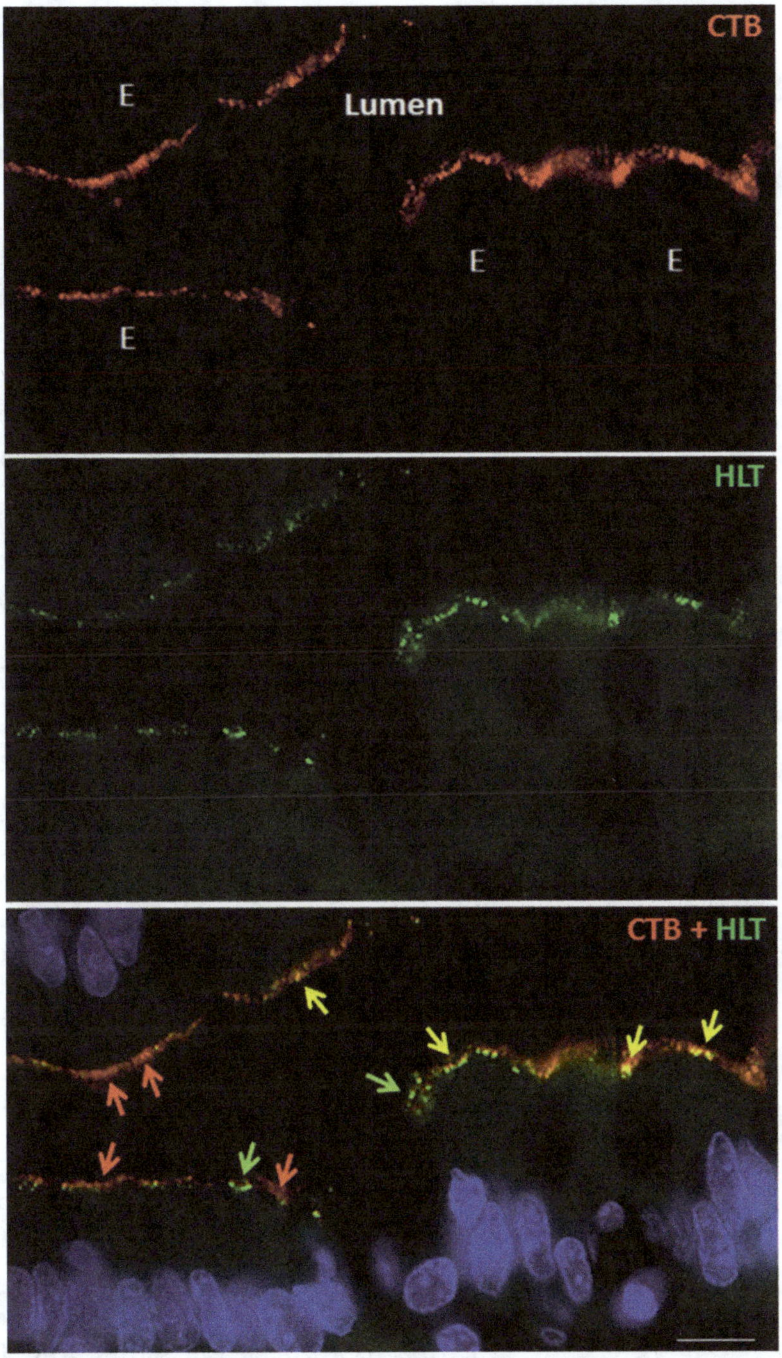

Fig. 1 Binding and uptake of CTB (cholera toxin B subunit) and HLT (heat-labile toxin of enterotoxigenic *E. coli*) in organ-cultured porcine jejunal explants. The mucosal explants were cultured for 1 h at 37 °C in RPMI medium in the presence of 10 μg/ml Alexa 594-conjugated CTB and 10 μg/ml FITC-conjugated HLT. Images of cryosections show binding of both enterotoxins to the luminal brush border of enterocytes (E) along the villi and uptake into subapical punctae. In some cells, only CTB (*red arrows*) or HLT (*green arrows*) was internalized,

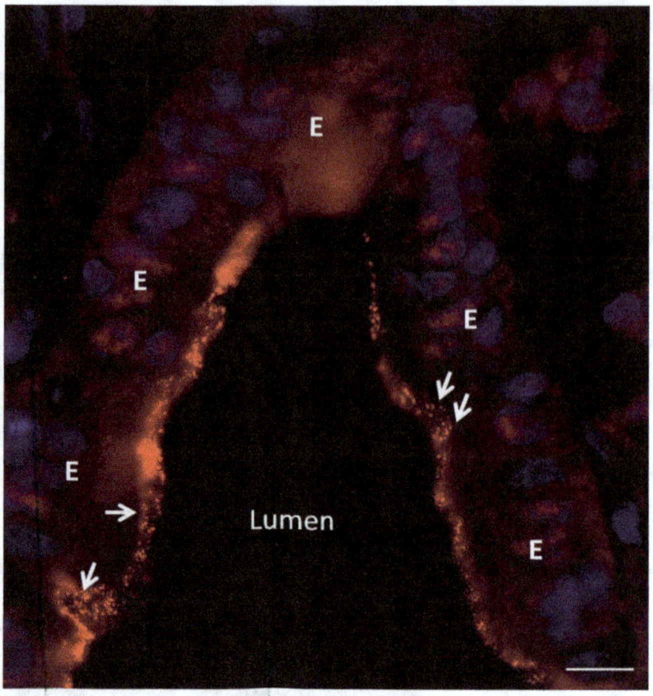

Fig. 2 Binding and uptake of SEB (*S. aureus* enterotoxin B) in organ-cultured porcine jejunal explants. Organ culture in the presence of 10 μg/ml FITC-SEB was performed as described in the legend to Fig. 1. After culture the explants were quickly rinsed in fresh medium and fixed overnight in 4 % paraformaldehyde in PBS. Cryosections of the explants were cut and immunolabeled with a mouse monoclonal antibody to fluorescein (Invitrogen), followed by labeling with Alexa 594-conjugated rabbit secondary antibodies (Invitrogen). Nuclei were visualized by DAPI. The image shows binding and uptake of SEB along the brush border of the enterocytes (E). Internalized SEB is seen as distinct punctae which in some cells (marked by *arrows*) had penetrated deep into the cytoplasm. Bar: 10 μm

2 Materials

1. Scalpels, scissors, and tweezers for excision and trimming of mucosal explants.
2. A suitable culture medium, for instance Roswell Park Memorial Institute (RPMI) medium (*see* **Note 1**).
3. 10 % (v/v) fetal calf serum, 100 U/ml of penicillin, and 0.1 mg/ml of streptomycin are recommended to be added to the culture medium for longer periods of culture (>5 h).
4. "Falcon" dishes for organ culture (Fig. 3) are in vitro fertilization dishes in polystyrene, 60 × 15 mm style with a center well, and a triangular grid of stainless steel (Cat. # 353653, Beckton Dickinson, www.bdbiosciences.com).
5. A suitable incubator.

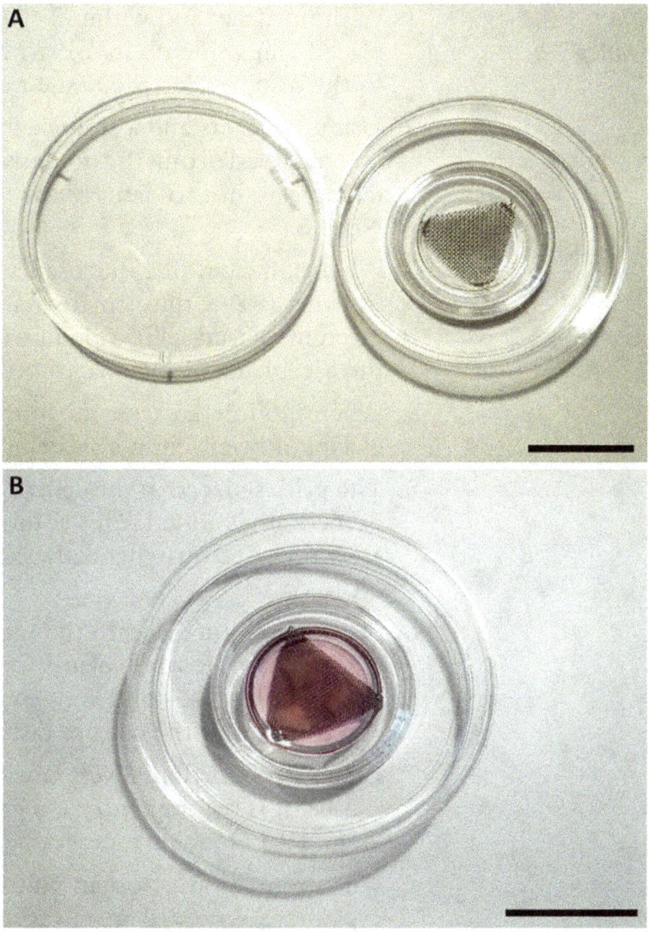

Fig. 3 A Falcon dish with lid for organ culture of intestinal explants. (**a**) Culture dish with a rimmed central well 2 cm in diameter in which a triangular grid of stainless steel is mounted. (**b**) Three mucosal explants placed on the metal grid villus side upwards and 1 ml of medium added to the central well. This volume ensures that the explants are just submerged in the medium. The medium should be replaced once for every 1–2 h of culture. Bars: 2 cm

3 Methods

3.1 Obtaining Intestinal Segments from a Donor Animal (See Note 2)

1. In a licensed facility for animal experimentation, a donor pig is placed at an operating table, maintained under anaesthesia, and connected to a respirator (*see* **Note 3**).
2. An abdominal incision is performed to expose the small intestine and to excise small segments (~10 cm) of the desired parts of the intestine (*see* **Note 4**).
3. The excised intestinal segments are cut open longitudinally to expose the mucosal surface before quickly being immersed in ice-cold RPMI medium.

3.2 Organ Culture Setup

1. Without delay and within 5–10 min, the intestinal segment is placed villus side up in a Petri dish. Medium is quickly added to the dish to keep the tissue moist at all times.
2. Using tweezers and a scalpel, small (~1 cm^2) pieces of mucosa are removed from the underlying serosal/muscularis layers, taking care not to damage the villus surface in the process (*see* **Note 5**).
3. When sufficient free mucosa has been obtained, small uniform explants (~2–3 mm squares) are excised with the scalpel and carefully placed villus side upwards on the metal grids. Up to three explants of this size can be mounted on each grid.
4. When mounted, excess medium is soaked off from the grid by placing it briefly atop a sheet of tissue paper.
5. The grid is placed in the central well of a culture dish and 1 ml of medium is added. This volume will ensure that the explants are immersed just beneath the surface of the medium (*see* **Note 6**).
6. Place the culture dish in the incubator at 37 °C. We routinely culture the mucosal explants for 15 min to allow for acclimatization before starting any experiment such as addition of enterotoxins to the culture dish (*see* **Note 7**).
7. Change the medium every hour. For longer periods of culture (>5 h), it is recommended to gas the cultured tissue with O_2/CO_2 (19:1) for short intervals (3–20 min) every 5 h [4, 5].
8. For termination of organ culture, the grid with the mucosal explants is removed from the culture dish with tweezers and washed once very gently with fresh medium by use of a Pasteur pipette.
9. Excess medium is soaked off as described in **step 4**. The further procedure depends upon which type of analysis will be performed (*see* **Note 8**).

4 Notes

1. The original medium recommended for organ culture was Trowell's T-8 medium [4], but to our knowledge this medium is no longer commercially available.
2. The protocol described here relates to our work using porcine small intestine. Organ culture of small intestine from smaller animals such as rabbit [10], guinea pig [11], rat [12], and mouse [13] has been reported by other investigators.

3. It is our experience that the tissue viability of intestinal mucosal explants crucially depends upon a maintained blood circulation in the donor animal at this step. Circulatory arrest prior to removal of the intestinal segment, even for a short period of time, is harmful to the integrity of the intestinal mucosa. For organ culture of intestine from smaller animals, such as mice, the intestine is removed immediately after sacrifice of the animal.
4. All animal handling steps must be performed by licensed staff and subject to national laws and regulations governing animal experimentation.
5. This maneuver can be tricky to perform and some practice may be required. Cutting/scraping with the scalpel at a fixed angle to the tissue with one hand, whilst holding the tissue firmly in place with the tweezers in the other hand, is one way to obtain satisfactory results. For organ culture of intestine from mice, we omit this step and instead excise small pieces of whole intestine to be placed directly in culture dishes.
6. Overall, the procedure described in points 2–4 should be completed within ~15 min to obtain a good viability during the subsequent organ culture. This sets a limit as to how many culture dishes that can be set up by one person in a single experiment; 10–12 culture dishes is a feasible number.
7. When adding small volumes of agents to the culture medium, the tip of the pipette should not be pointed directly at the tissue. After addition, the medium is mixed by repeated, gentle use of a 200 μl pipette, still taking care not to point it at the tissue.
8. If biochemical analyses are to be performed, the tissue is quickly transferred to an Eppendorf tube and frozen until further processing. For microscopy, the explants are directly immersed in a suitable fixative, for instance 4 % paraformaldehyde in PBS, pH 7.4, and fixed at 4 °C for 2 h or overnight. After fixation, the explants are kept in 1 % paraformaldehyde in PBS, pH 7.4 at 4 °C until further processing.

Acknowledgements

The work was supported by grants from Augustinus Fonden, Aase og Ejnar Danielsens Fond, Brødrene Hartmanns Fond, Fonden til Lægevidenskabens Fremme, and Hørslev Fonden. The funders had no role in study design, data collection and analysis, decision to publish, or preparation of the manuscript.

References

1. Guerrant RL, Oria R, Bushen OY, Patrick PD, Houpt E, Lima AA (2005) Global impact of diarrheal diseases that are sampled by travelers: the rest of the hippopotamus. Clin Infect Dis 41(Suppl 8):S524–S530
2. Glenn GM, Francis DH, Danielsen EM (2009) Toxin-mediated effects on the innate mucosal defenses: implications for enteric vaccines. Infect Immun 77:5206–5215
3. Choi YW, Kotzin B, Herron L, Callahan J, Marrack P, Kappler J (1989) Interaction of Staphylococcus aureus toxin "superantigens" with human T cells. Proc Natl Acad Sci U S A 86:8941–8945
4. Browning TH, Trier JS (1969) Organ culture of mucosal biopsies of human small intestine. J Clin Invest 48:1423–1432
5. Danielsen EM, Sjostrom H, Noren O, Bro B, Dabelsteen E (1982) Biosynthesis of intestinal microvillar proteins. Characterization of intestinal explants in organ culture and evidence for the existence of pro-forms of the microvillar enzymes. Biochem J 202:647–654
6. Hansen GH, Niels-Christiansen LL, Immerdal L, Hunziker W, Kenny AJ, Danielsen EM (1999) Transcytosis of immunoglobulin A in the mouse enterocyte occurs through glycolipid raft- and rab17-containing compartments. Gastroenterology 116:610–622
7. Hansen GH, Niels-Christiansen LL, Immerdal L, Danielsen EM (2006) Antibodies in the small intestine: mucosal synthesis and deposition of anti-glycosyl IgA, IgM, and IgG in the enterocyte brush border. Am. J. Physiol Gastrointest. Liver Physiol 291:G82–G90
8. Danielsen EM, Hansen GH, Karlsdottir E (2013) Staphylococcus aureus enterotoxins A- and B: binding to the enterocyte brush border and uptake by perturbation of the apical endocytic membrane traffic. Histochem Cell Biol 139:513–524
9. Hansen GH, Dalskov SM, Rasmussen CR, Immerdal L, Niels-Christiansen LL, Danielsen EM (2005) Cholera Toxin Entry into Pig Enterocytes Occurs via a Lipid Raft- and Clathrin-Dependent Mechanism. Biochemistry 44:873–882
10. Kagnoff MF, Donaldson RM Jr, Trier JS (1972) Organ culture of rabbit small intestine: prolonged in vitro steady state protein synthesis and secretion and secretory IgA secretion. Gastroenterology 63:541–551
11. Kedinger M, Haffen K, Hugon JS (1974) Organ culture of adult guinea-pig intestine. I. Ultrastructural aspect after 24 and 48 hours of culture. ZZellforsch Mikrosk Anat 147:169–181
12. Shields HM, Yedlin ST, Bair FA, Goodwin CL, Alpers DH (1979) Successful maintenance of suckling rat ileum in organ culture. Am J Anat 155:375–389
13. Berteloot A, Chabot JG, Menard D, Hugon JS (1979) Organ culture of adult mouse intestine. III. Behavior of the proteins, DNA content and brush border membrane enzymatic activities. In Vitro 15:294–299

Chapter 15

Quantification of a Selective Expansion of T Cell Receptor Vβ by Superantigen Using Real-Time PCR

Joo Youn Park and Keun Seok Seo

Abstract

Selective expansion of T cells bearing specific T cell receptor Vβ segments is a hallmark of superantigens. Analyzing Vβ specificity of superantigens is important for characterizing newly discovered superantigens and understanding differential T cell responses to each toxin. Here, we describe a real-time PCR method using SYBR green I and primers specific to Cβ and Vβ genes for an absolute quantification. The established method was applied to quantify a selective expansion of T cell receptor Vβ expansion by superantigens and generated accurate, reproducible, and comparable results.

Key words Superantigen, T cell receptor Vβ, Quantitative real-time PCR

1 Introduction

The T cell receptor (TCR) is an α and β chain heterodimer that recognizes antigen-derived peptides presented by the major histocompatibility complex II (MHC II) molecules on antigen-presenting cells, thereby triggering a clonal expansion of T cells [1]. The TCR consists of a variable and constant region, where the variable region interacts with peptide and MHC II, thus determining the antigen specificity of T cells [2, 3]. A high diversity of the variable region of TCR is generated by a somatic recombination of variable (V), diversity (D), and joining (J) genes with a constant (C) gene and imprecise joining of VDJ segments during thymic development [4]. Cloning and sequencing human Vβ genes have identified 49 functional Vβ subgroup genes. Based on the sequence similarity, these genes were grouped into 24 different Vβ groups. Due to the sequence variations, some Vβ groups contain multiple subgroups [2, 5, 6].

Staphylococcal enterotoxins (SEs), SE-like toxin (SEl), and toxic shock syndrome toxin-1 (TSST-1) are prototypical microbial superantigens (SAg). Thus far, 23 SEs and SEls including SEA through X, excluding F, and TSST-1 have been characterized in *Staphylococcus aureus* [7]. Unlike conventional antigens, most SAgs

Amanda J. Brosnahan (ed.), *Superantigens: Methods and Protocols*, Methods in Molecular Biology, vol. 1396,
DOI 10.1007/978-1-4939-3344-0_15,

directly bind to the specific variable regions of TCR β chain (Vβ) and outside of the peptide-binding groove of MHC II. This binding triggers a clonal expansion of T cells bearing specific TCR Vβ segments, leading to a massive production of proinflammatory cytokines and chemokines [8]. Thus, SAgs induce antigen-independent, Vβ-dependent T cell proliferation. The shared biological properties of SAg such as Vβ specificity and cytokine responses were closely correlated with the amino acid sequence and 3-dimensional structure of SAgs. Therefore, the analysis of Vβ specificity of SAg is important for characterizing newly discovered SAg and understanding T cell responses to the SAg.

Several approaches are used to analyze the Vβ specificity of SAg including northern blotting, semiquantitative PCR [9], and flow cytometry using monoclonal antibodies specific to Vβ segments [10, 11]. Shortcomings such as a lack of available reagents limit the practical application of these approaches. We developed a quantitative real-time PCR (qRT-PCR) method using SYBR green I and primers specific to 22 Vβ groups and constant β chain (Cβ) genes [12]. The specificity of established method was verified by sequencing PCR amplification products showing that 36 out of 49 functional Vβ subgroup genes were successfully amplified. Standard curves for primers to Cβ and Vβ genes were generated to allow an absolute quantification. The established method was applied to assess the Vβ specificity of SAgs and showed reproducible and comparable results from previous approaches.

2 Materials

2.1 Preparation and Stimulation of Enriched Human Lymphocytes

1. BD Vacutainer Saftey-Lok Blood collection set.
2. Blood collection tube containing heparin.
3. Phosphate-buffered saline (PBS).
4. Ficoll-Hypaque plus solution (density 1.077 g/l).
5. Hanks' balanced salt solution (HBSS).
6. Complete RPMI1640 medium: RPMI medium supplemented with 2 % FBS, 100 U penicillin G, and 100 μg/ml streptomycin.
7. Beckman GPR centrifuge with GH-3.7 horizontal rotor (or equivalent temperature-controlled centrifuge).
8. Endotoxin-free SAg.
9. Murine monoclonal antibody (mAb) specific to human CD3 (OKT3, Sigma-Aldrich).

2.2 RNA Extraction

1. RNeasy Mini Kit (Qiagen).
2. RNase free TURBO DNase I (Ambion).
3. 70 % Ethanol.

2.3 cDNA Synthesis

1. DNA, RNA-free 0.2 ml PCR tube.
2. Thermocycler.
3. dNTP mix (10 mM each, Life Technologies).
4. oligo(dT)$_{20}$ (0.5 μg/μl, Life Technologies).
5. Superscriptase III first strand synthesis kit (Life Technologies).

2.4 Quantitative Real-Time PCR

1. Applied Biosystems 7500 real-time PCR system or equivalent real-time PCR thermal cycler.
2. MicroAmp Optical 96-well reaction plate (Life Technologies).
3. MicroAmp Optical plate seals (Life Technologies).
4. Power SYBR Green PCR Master mix (Life Technologies).
5. Primers specific to the Cβ and Vβ genes.
6. PCR purification kit (Qiagen).
7. pCR2.1 plasmid vector (Life Technologies).
8. Plasmid MiniPrep Kit (Qiagen).

3 Methods

3.1 Preparation and Stimulation of Enriched Human Lymphocytes

1. Collect whole blood (10–20 ml) from the healthy donor by venipuncture using BD Vacutainer Saftey-Lok Blood collection set (21G needle) and Blood collection tube containing heparin (14 U/ml blood).
2. Transfer whole blood into the 50 ml conical polypropylene tube and add an equal volume of PBS. Mix well.
3. Slowly overlay the diluted blood onto the Ficoll-Hypaque plus solution. Use 1 ml Ficoll-Hypaque per 3 ml blood diluted with PBS.
4. Centrifuge at 2000 rpm (900 × *g*), 18–20 °C, for 30 min without brake.
5. Using sterile pipet, carefully remove the upper layer containing the plasma and platelets and discard. Using another pipet, carefully transfer buffy coat that contains the mononuclear cells to the 50 ml conical polypropylene tube.
6. Wash mononuclear cells three times by adding 45 ml of HBSS and centrifuging at 1300 rpm (400 × *g*), 18–20 °C, for 10 min (*see* **Note 1**).
7. Resuspend the cells in complete RPMI medium and culture in cell culture Petri dishes overnight at 37 °C and in 5 % CO_2.
8. The following day, non-adherent, lymphocyte-enriched cells are collected, washed, and resuspended in complete RPMI medium at a final concentration of 2.5×10^6 cells/ml.

9. Each SAg (0.5 μg/ml) is added to lymphocyte-enriched cells. Cells are cultured for 4 days (37 °C, 5 % CO_2) and harvested by centrifuging at 1300 rpm (400 × *g*), 4 °C, for 10 min.

3.2 Total RNA Extraction

1. Approximately 5×10^6 cells are resuspended in 350 μl of buffer RLT and homogenized by passing the lysate ten times through a 20-gauge needle (*see* **Note 2**).
2. Centrifuge the lysate at 12,000 × *g* for 3 min and carefully transfer the supernatant to a new microcentrifuge tube.
3. Add an equal volume of 70 % ethanol and mix by pipetting.
4. Transfer the mixture to the RNeasy spin column, centrifuge at 8000 × *g* for 15 s, and discard the flow through.
5. Add 700 μl Buffer RW1 to the RNeasy spin column, centrifuge at 8000 × *g* for 15 s, and discard the flow through.
6. Add 500 μl Buffer RPE to the RNeasy spin column, centrifuge at 8000 × *g* for 15 s, and discard the flow through.
7. Add 500 μl Buffer RPE to the RNeasy spin column, centrifuge at 8000 × *g* for 2 min, and discard the flow through.
8. Place the RNeasy spin column in a new microcentrifuge tube and add 50 μl RNase-free water directly to the spin column membrane. Centrifuge at 8000 × *g* for 1 min.
9. Add RNase-free water to make the volume up to 89 μl.
10. Add 10 μl 10× TURBO DNase Buffer and 1 μl TURBO DNase (2 U/μl)
11. Incubate at 37 °C for 30 min.
12. Add 350 μl Buffer RLT and mix well by pipetting.
13. Add 250 μl 100 % ethanol and mix well by pipetting.
14. Transfer the mixture to a new RNeasy spin column, centrifuge at 8000 × *g* for 15 s, and discard the flow through.
15. Add 500 μl Buffer RPE to the RNeasy spin column, centrifuge at 8000 × *g* for 15 s, and discard the flow through.
16. Add 500 μl Buffer RPE to the RNeasy spin column, and centrifuge at 8000 × *g* for 2 min, and discard the flow through.
17. Place the RNeasy spin column in a new microcentrifuge tube and add 50 μl RNase-free water directly to the spin column membrane. Centrifuge at 8000 × *g* for 1 min.
18. Determine the quantity and quality of RNA using Nanodrop and adjust the quantity to 1 μg/5 μl (*see* **Note 3**).

3.3 cDNA Synthesis

1. Add the following in a 0.2 ml PCR tube:
   - 5 μl total RNA (1 μg)
   - 1 μl dNTP mix (10 mM each dATP, dGTP, dCTP, and dTTP)
   - 1 μl oligo$(dT)_{20}$ (500 ng)

- 7 μl RNase-free water
2. Incubate for 5 min at 65 °C, and then place on ice for 2 min.
3. Add the following cDNA synthesis mix to the tube:
   - 4 μl 5× Reverse transcriptase buffer
   - 1 μl 0.M DTT
   - 1 μl Superscriptase III RT (200 U/μl)
4. Incubate for 50 min at 50 °C.
5. Incubate for 5 min at 85 °C to terminate the reactions, and then place on ice.
6. Dilute cDNA with 980 μl DNase-free water and store at −80 °C until used.

3.4 Primer Design

Primers specific to the Cβ and 23 different Vβ groups were designed using Primer Express version 2.0 (Applied Biosystems). Some Vβ groups have multiple Vβ subgroup genes showing high sequence similarity. Therefore, some Vβ primers are expected to amplify multiple Vβ subgroup genes within the corresponding Vβ group. Sequencing analysis of PCR products generated using these primers showed that 36 out of 49 functional Vβ genes were amplified. The primer sequences and amplified Vβ subgroup genes are summarized in Table 1.

3.5 Quantitative Real-Time PCR

For absolute quantification of the Cβ and Vβ genes, the standard curves for primers to Cβ and Vβ genes were generated.

3.5.1 Standard Curve

1. The Cβ and Vβ genes were amplified by PCR using primers listed in Table 1. Each PCR product was purified using a PCR purification kit and then cloned into pCR2.1 plasmid vector. Cloned pCR2.1 plasmid vectors were purified using a plasmid MiniPrep kit and the concentration of plasmid was determined by measuring the absorbance at 260 nm using a Nanodrop. The plasmid copy number was determined by the following formula [13, 14]:

$$\text{Number of plasmid copies (molecules)} = \frac{X\ \text{ng} \times 6.0221 \times 10^{23}\ \text{molecules / mol}}{(N \times 660\ \text{g / mol}) \times 1 \times 10^{9}\ \text{ng / g}}$$

where

X = the amount of plasmid (ng)

N = the size (base pair) of plasmid

2. Set up triplicate qRT-PCR reactions (25 μl each) consisting of:
   - 12.5 μl Power SYBR Green PCR Master mix
   - 5 μl (2 pmol) forward and reverse primers (2.5 μl each)
   - 5 μl tenfold serially diluted plasmid template ($2.5 - 2.5 \times 10^5$ copies)
   - 2.5 μl DNase-free water

Table 1
List of primers[a] and amplified Vβ gene(s), reproduced from [12] with permission from JTM

Primer name	GenBank access number	Forward primer ('5 to 3')	Reverse primer ('5 to 3')	Amplified Vβ gene(s)[b]
Cβ	L36092	tccagttctacgggctctcg	gacgatctgggtgacgggt	
VB1	L36092	ggagcaggcccagtggat	cgctgtccagttgctggtat	TCRVB1s1
VB2	M11955	gagtctcatgctgatggcaact	tctcgacgccttgctcgtat	TCRVB2s1
VB3	U08314	tcctctgtcgtgtggccttt	tctcgagctctgggttactttca	TCRVB3s1
VB4	L36092	ggctctgaggccacatatgag	ttaggtttgggcggctgat	TCRVB4s1
VB5	L36092	gctccaggctgctctgttg	tttgagtgactccagcctttactg	TCRVB5s1, 5 s3
VB6	X61440	ggcagggcccagagtttc	gggcagccctgagtcatct	TCRVB6s1, 6 s2, 6 s3, 6 s4, 6 s5, 6 s6
VB7	U07977	aagtgtgccaagtcgcttctc	tgcagggcgtgtaggtgaa	TCRVB7s1, 7 s2, 7 s3
VB8	X07192	tgcccgaggatcgattctc	tctgagggctggatcttcaga	TCRVB8s1, 8 s2, 8 s3
VB9	U07977	tgcccgaggatcgattctc	tctgagggctggatcttcaga	TCRVB9s1
VB11	L36092	catctaccagaccccaagatacct	atggcccatggtttgagaac	TCRVB11s1
VB12	U03115	gttcttctatgtggccctttgtct	tcttgggctctgggtgattc	TCRVB12s1, 12 s3
VB13[c]	L36092	tggtgctggtatcactgaccaa	ggaaatcctctgtggttgatctg	TCRVB13s1, 13 s6
VB13[d]	X61445	tgtgggcaggtccagtga	tgtcttcaggacccggaatt	TCRVB13s2, 13 s9
VB14	L36092	gctccttggctatgtggtcc	ttgggttctgggtcacttgg	TCRVB14s1
VB15	M11951	tgttacccagaccccaagga	tgacccttagtctgagaacattcca	TCRVB15s1
VB16	X06154	cggtatgcccaacaatcgat	caggctgcaccttcagagtaga	TCRVB16s1
VB17	U48260	caaccaggtgctctgctgtgt	gactgagtgattccaccatcca	TCRVB17s1

VB18	L36092	ggaatgccaaaggaacgattt	tgctggatcctcaggatgct	TCRVB18s1
VB20	L36092	aggtgccccagaatctctca	ggagcttcttagaactcaggatgaa	TCRVB20s1
VB21	M33233	gctgtggctttttggtgtga	caggatctgccggtaccagta	TCRVB21s1
VB22	L36092	tgaaagcaggactcacagaacct	tcacttcctgtcccatctgtgt	TCRVB22s1
VB23	U03115	ttcagtggctgctggagtca	cagagtggctgtttccctcttt	TCRVB23s1
VB24	U03115	acccctgataacttccaatcca	cctggtgagcggatgtcaa	TCRVB24s1

[a]The pseudogenes (Vβ10 and Vβ19) were not included
[b]Vβ subgroup nomenclature followed the classification of Arden et al. [6]
[c]VB13A corresponds to Vβ13.1 in previous studies
[d]VB13B corresponds to Vβ13.2 in previous studies

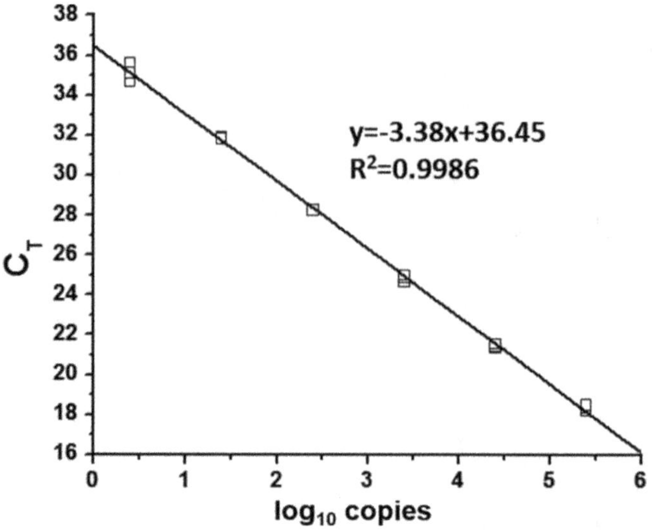

Fig. 1 A representative standard curve generated for the Cβ gene, reproduced from [12] with permission from JTM. Quantitative real-time PCR was performed using tenfold serially diluted templates ($2.5–2.5 \times 10^5$ copies). The C_T was determined and plotted over the $\log_{10}$ copies to calculate the slope, Y axis intercept, and correlation coefficiency (R^2)

3. Thermocycler conditions consisted of an initial denaturation at 95 °C for 10 min, followed by 40 cycles of a denaturation at 95 °C for 15 s and an extension at 60 °C for 1 min. Fluorescent data is measured during an extension period. After 40 cycles, a melting curve analysis is performed. The base line and threshold cycle (C_T) are determined using the Sequence Detector Systems version 1.2.2 (Applied Biosystems).
4. The standard curve is generated by plotting the C_T vs. the $\log_{10}$ copies of serially diluted plasmid harboring Cβ and Vβ genes ($2.5 - 2.5 \times 10^5$ copies). The slope, intercept, and correlation coefficiency (R^2) are determined using Microcal OriginPro Vesion 7.5 (OriginLab) (*see* **Note 4**).

   Representative standard curve for the Cβ gene is shown in Fig. 1. Standard curves for the Cβ and Vβ genes are summarized in Table 2.

3.5.2 Quantification of Vβ Gene

1. Set up triplicate qRT-PCR reactions (25 μl) consisted of:
   - 12.5 μl Power SYBR Green PCR Master mix
   - 5 μl 2 pmol forward and reverse primers (2.5 μl each) specific to the Cβ or Vβ gene
   - 5 μl cDNA prepared in Subheading 3.3
   - 2.5 μl DNase-free water

Table 2
Standard curve slopes, *Y*-axis intercepts, and correlation coefficients (R^2), reproduced from [12] with permission from JTM

Primers	Slope	*Y*-axis intercept	Correlation coefficient (R^2)
Cβ	-3.38	36.45	0.9986
VB1	-3.39	36.54	0.9977
VB2	-3.36	36.38	0.9982
VB3	-3.41	36.57	0.9987
VB4	-3.37	36.62	0.9984
VB5	-3.35	36.33	0.9976
VB6	-3.40	36.53	0.9978
VB7	-3.36	36.43	0.9983
VB8	-3.37	36.40	0.9986
VB9	-3.38	36.49	0.9985
VB11	-3.41	36.52	0.9986
VB12	-3.42	36.53	0.9972
VB13A	-3.34	36.34	0.9978
VB13B	-3.41	36.54	0.9974
VB14	-3.36	36.33	0.9981
VB15	-3.35	36.44	0.9976
VB16	-3.37	36.44	0.9984
VB17	-3.39	36.53	0.9982
VB18	-3.35	36.44	0.9986
VB20	-3.33	36.39	0.9973
VB21	-3.36	36.38	0.9986
VB22	-3.39	36.47	0.9981
VB23	-3.37	36.43	0.9980
VB24	-3.41	36.53	0.9984

2. Perform and analyze the qRT-PCR as described in Subheading 3.5.1 (*see* **Note 5**).
3. For absolute quantification, the absolute copy numbers of the Cβ or Vβ gene are calculated by extrapolating the C_T to the standard curve listed in Table 2 (*see* **Note 6**).

4. The percentage of each Vβ (%Vβ) is calculated by the following equation:

$$\%V\beta_n = \left(\frac{V\beta_n}{C\beta}\right) \times 100$$

where
$\%V\beta_n$ = the percentage of indicated (n) Vβ group
$V\beta_n$ = the copy number of indicated (n) Vβ group
$C\beta$ = the copy number of Cβ gene

5. Selective expansion of Vβ by SAg is determined when the %Vβ*n* from the culture stimulated with SAg is significantly higher than the corresponding %Vβ*n* from the unstimulated culture (without stimuli) by paired *t*-test ($p < 0.01$).

   The established method was applied to determine the Vβ specificity of SEA, SEB, SEC1, SED, SEE, SEG, SEI, SElM, SElN, SElO, and TSST-1. Results are summarized in Fig. 2 and Table 3 and showed accurate, reproducible, and comparable results observed in previous studies [15–17].

4 Notes

1. If the cell pellet is contaminated with red blood cells, resuspend the cell pellet in 1×ACK lysis buffer (Life Technologies, add 1 ml 1 × ACK lysis buffer per 2 ml of original blood volume), incubate at room temperature for 5 min, and add equal volume of PBS.
2. The cell pellet or lysate can be stored in −80°C for a month. Frozen samples should be completely thawed and continue with **step 3**.
3. Normally, we do not experience DNA contamination in the RNA preparation after the DNase I treatment. A contamination of DNA in RNA preparation could be verified by PCR reactions with primers specific to GAPDH (forward: 5′-GCAAATTCCATGGCACCGT-3′; reverse: 5′-TCGCCCCACTTGATTTTGG-3′).
4. Standard curves generated for Cβ and Vβ genes using other real-time PCR equipment such as iCycler (Bio-rad) were not significantly different from those generated by ABI 7500 real-time PCR systems.
5. Melting curve analysis of primers for Vβ7, 12, 13A, and 17 may show multiple peaks due to the amplification of multiple Vβ subgroup genes that have heterogeneity in melting temperature. Although primers for Vβ5, 6 13B, and 21 also amplify multiple Vβ subgroup genes, melting curve analysis showed a single peak due to the homogeneity in melting temperatures.

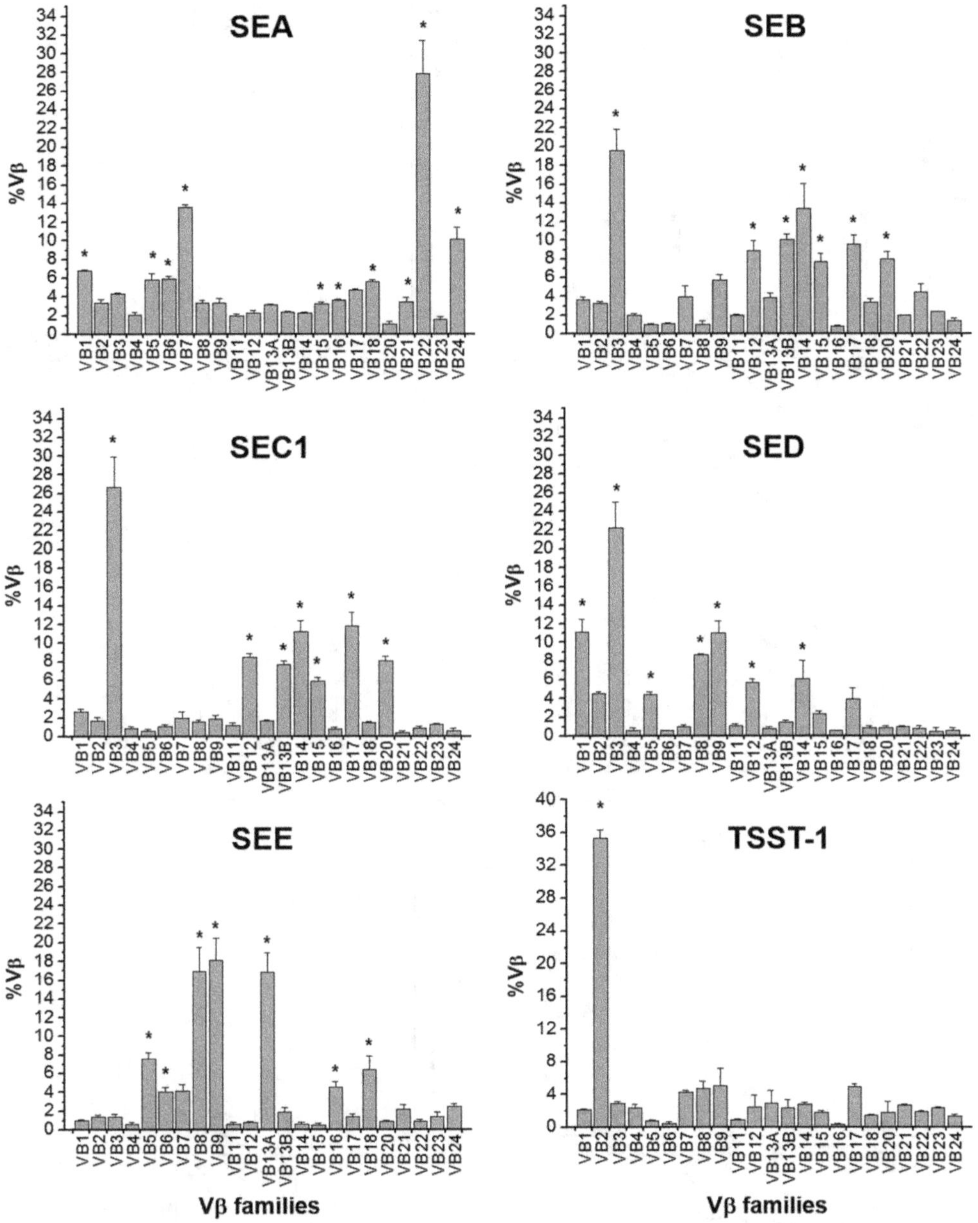

Fig. 2 Selective expansion of TCR Vβ by SAgs, reproduced from [12] with permission from JTM. Enriched human lymphocytes were stimulated with an indicated SAg (0.5 μg/ml) for 4 days. Quantitative real-time PCR was performed and the %Vβ was calculated. Selective expansion of Vβ by SAg is determined when the %Vβ*n* from the culture stimulated with SAg is significantly higher than the corresponding %Vβ*n* from the unstimulated culture (without stimuli) by paired *t*-test. The asterisk indicates a statistically significance ($p < 0.01$)

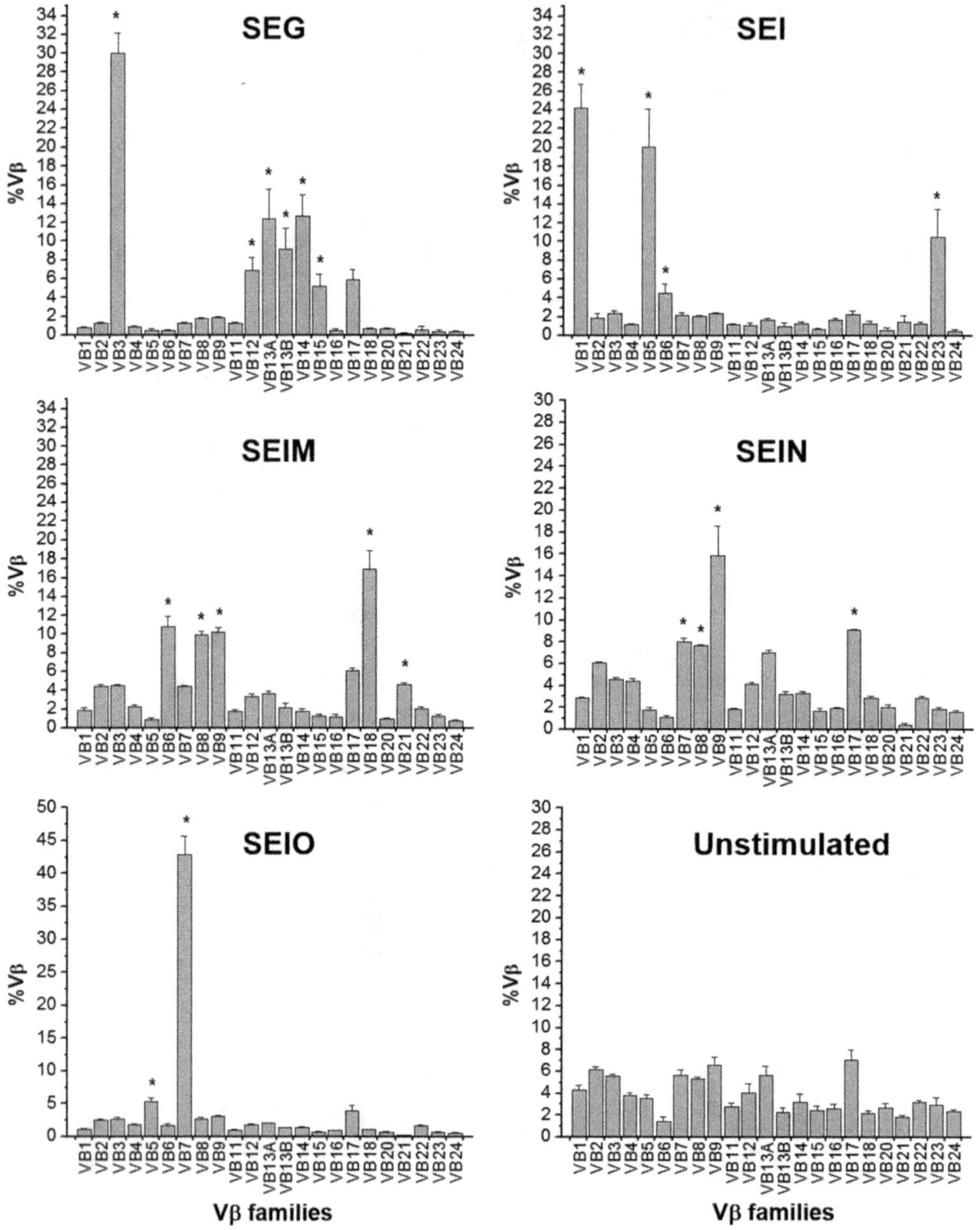

Fig. 2 (continued)

Table 3
Summary of Vβ specificity observed with these methods and comparison with those in selected previous studies, reproduced from [12] with permission from JTM

SAgs	Vβ specificity observed in this study	Vβ specificity observed in previous studies[a]	References
SEA	Vβ1, 5, 6, 7, 15, 16, 18, 21, 22, 24	Vβ1, 5, 6, 7, 9, 16, 18, 21	[15]
SEB	Vβ3, 12, 13B[b], 14, 15, 17, 20	Vβ1, 3, 6, 12, 13.2, 15, 17, 20	[16]
SEC1	Vβ3, 12, 13B, 14, 15, 17, 20	Vβ3, 12, 13.2, 14, 15, 17, 20	[9]
SED	Vβ1, 3, 5, 8, 9, 12, 14	Vβ1, 5, 6, 7, 8, 12	[7, 18]
SEE	Vβ5, 6, 8, 9, 13A[c], 16, 18	Vβ5, 6, 8, 13.1, 18, 21	[15, 16]
SEG	Vβ3, 12, 13A, 13B, 14, 15	Vβ3, 12, 13, 14	[17]
SEI	Vβ1, 5, 6, 23	Vβ1, 5, 6, 23	[17]
SEIM	Vβ6, 8, 9, 18, 21	Vβ6, 8, 9, 18, 21	[17]
SEIN	Vβ7, 8, 9, 17	Vβ9	[17]
SEIO	Vβ5, 7	Vβ5, 7, 22	[17]
TSST-1	Vβ2	Vβ2	[16]

[a]Vβ specificities were results from previous studies using semiquantitative PCR or FACS methods
[b]Vβ13B corresponds to Vβ13.2 in previous studies
[c]Vβ13A corresponds to Vβ13.1 in previous studies

6. To synchronize real-time PCR data analysis, the base line signal and threshold were set to be determined automatically by the Sequence Detector System version 1.2.2.

References

1. Davis MM, Boniface JJ, Reich Z, Lyons D, Hampl J, Arden B, Chien Y (1998) Ligand recognition by alpha beta T cell receptors. Annu Rev Immunol 16:523–544
2. Rowen L, Koop BF, Hood L (1996) The complete 685-kilobase DNA sequence of the human beta T cell receptor locus. Science 272:1755–1762
3. Davis MM, Bjorkman PJ (1988) T-cell antigen receptor genes and T-cell recognition. Nature 334:395–402
4. Behlke MA, Spinella DG, Chou HS, Sha W, Hartl DL, Loh DY (1985) T-cell receptor beta-chain expression: dependence on relatively few variable region genes. Science 229:566–570
5. Lefranc M, Lefranc G (2001) The T cell receptor. Academic, New York
6. Arden B, Clark SP, Kabelitz D, Mak TW (1995) Human T-cell receptor variable gene segment families. Immunogenetics 42:455–500
7. Seo KS, Bohach GA (2007) Staphylcoccus aureus. In: Doyle MM, Beucaht LR (eds) Food microbiology: fundamentals and Frontiers, 3rd edn. ASM Press, Washington, DC, pp 493–518
8. Bohach GA (2006) *Staphylococcus aureus* Exotoxins. In: Fischetti VA, Novick RP, Ferretti JJ, Portnoy DA, Rood JI (eds) Gram-positive pathogens, 2nd edn. ASM Press, Washington, DC, pp 464–477
9. Deringer JR, Ely RJ, Stauffacher CV, Bohach GA (1996) Subtype-specific interactions of type C staphylococcal enterotoxins with the T-cell receptor. Mol Microbiol 22:523–534
10. Pilch H, Hohn H, Freitag K, Neukirch C, Necker A, Haddad P, Tanner B, Knapstein PG, Maeurer MJ (2002) Improved assessment of T-cell receptor (TCR) VB repertoire in clinical

specimens: combination of TCR-CDR3 spectratyping with flow cytometry-based TCR VB frequency analysis. Clin Diagn Lab Immunol 9:257–266

11. Bercovici N, Duffour MT, Agrawal S, Salcedo M, Abastado JP (2000) New methods for assessing T-cell responses. Clin Diagn Lab Immunol 7:859–864
12. Seo KS, Park JY, Terman DS, Bohach GA (2010) A quantitative real time PCR method to analyze T cell receptor Vbeta subgroup expansion by staphylococcal superantigens. J Transl Med 8:2. doi:10.1186/1479-5876-8-2
13. Sambrook J, Fritsch EF, Maniatis T (1989) Concentration of DNA solution. In: Nolan C (ed) Molecular cloning: a laboratory manual, 2nd edn. Cold Spring Harbor Laboratory Press, New York, p Appendix C1
14. Yin JL, Shackel NA, Zekry A, McGuinness PH, Richards C, Putten KV, McCaughan GW, Eris JM, Bishop GA (2001) Real-time reverse transcriptase-polymerase chain reaction (RT-PCR) for measurement of cytokine and growth factor mRNA expression with fluorogenic probes or SYBR Green I. Immunol Cell Biol 79:213–221
15. Lamphear JG, Mollick JA, Reda KB, Rich RR (1996) Residues near the amino and carboxyl termini of staphylococcal enterotoxin E independently mediate TCR V beta-specific interactions. J Immunol 156:2178–2185
16. Choi YW, Kotzin B, Herron L, Callahan J, Marrack P, Kappler J (1989) Interaction of *Staphylococcus aureus* toxin "superantigens" with human T cells. Proc Natl Acad Sci U S A 86:8941–8945
17. Jarraud S, Peyrat MA, Lim A, Tristan A, Bes M, Mougel C, Etienne J, Vandenesch F, Bonneville M, Lina G (2001) *egc*, a highly prevalent operon of enterotoxin gene, forms a putative nursery of superantigens in *Staphylococcus aureus*. J Immunol 166:669–677
18. Kappler J, Kotzin B, Herron L, Gelfand EW, Bigler RD, Boylston A, Carrel S, Posnett DN, Choi Y, Marrack P (1989) V beta-specific stimulation of human T cells by staphylococcal toxins. Science 244:811–813

Chapter 16

Induction of Human Regulatory T Cells with Bacterial Superantigens

Stefano Caserta, Amanda L. Taylor, Nadia Terrazzini, and Martin J. Llewelyn

Abstract

Regulatory T cells (Tregs) that suppress the activation of immune effector cells limit immunopathology and are fast emerging as therapeutic targets for autoimmune and cancer disease. Tools enabling Treg in vitro-induction, expansion, and characterization and manipulation will help future clinical developments. In this chapter, *we describe in detail* how to use bacterial superantigens to induce human Tregs efficiently from peripheral blood mononuclear cells. How to assess human Treg phenotype and suppressive capacity are also described. *Technical details, variations, and alternative experimental conditions are provided.*

Key words Superantigen, Regulatory T cells, Human, Superantigen stimulation, Treg phenotype, Suppression assay, Cell purification, PBMC, Immunological methods

1 Introduction

Regulatory T cells (Tregs) counterbalance the activation of immune effector cells, which would otherwise promote overt immunopathology [1]. Based on their origin, Tregs are classified either as thymic-derived/natural Tregs (nTregs) or as peripheral/inducible Tregs (iTregs) [2]. Recent data suggest that nTregs and iTregs differ in their TCR usage [3]. The fate of Tregs is likely to be determined by antigen-(Ag)-specificity (including self- and pathogen-derived Ag) in a poorly defined manner [3–5]. Tregs are characterized by the expression of specific surface markers (such as CD25, GITR, CTLA-4, CD39, and others [5]) and the constitutive expression of the transcription factor FoxP3 [6, 7], although in humans FoxP3 expression on iTregs is unstable [8]. Other proteins such as Helios [9] and Neuropilin 1 [10, 11] have been suggested to mark Treg subsets, yet not univocally [12].

Amanda J. Brosnahan (ed.), *Superantigens: Methods and Protocols*, Methods in Molecular Biology, vol. 1396,
DOI 10.1007/978-1-4939-3344-0_16, © Springer Science+Business Media New York 2016

In the past 10 years, several methods have been established to induce and/or expand Tregs in vitro using either polyclonal or Ag-specific stimulation, in particular: (i) anti-CD3 with or without anti-CD28 stimulation in the presence of TGF-β1 and/or high doses of recombinant IL-2; a variant of these protocols also uses drugs such as retinoic acid to favor the conversion and/or expansion of Tregs from naïve T cell precursors [13, 14]; (ii) mixed lymphocyte reaction of host Tregs and allogeneic-donor irradiated APC supplemented with IL-2 [15]; and (iii) cultures in the presence of specific Ag (derived from pathogens or tumors) and high dose of IL-2 [16]. Protocols may differ dependent on the animal species used in the study (mice *versus* humans) and may yield to Treg expansion accompanied by effector T cell contamination, hence requiring a final step of Treg purification.

We previously showed that T cell stimulation with pathogen-derived toxins called superantigens results in expansion of human Treg populations which specifically express certain TCR Vβ-clonotypes [17]. Further, different superantigens selectively elicit the expansion of different TCR Vβ-clonotypes [18, 19]. We describe here the protocol that we used to generate superantigen-induced Tregs from human peripheral blood mononuclear cells. This method focuses primarily on the streptococcal pyrogenic exotoxin K/L (SPE-K/L) superantigen promoting the expansion of Vβ1$^+$ Tregs [17], however, it can be used for the induction of Tregs following exposure to other superantigens, including SPEA, staphylococcal enterotoxin (SE) A, and toxic shock syndrome toxin (TSST)-1 [17, 19]. Figure 1 shows the experimental outline for this method. Briefly, at day 1, blood is recovered from donors and the peripheral blood mononuclear cells (PBMCs) isolated. Unfractionated PBMCs or isolated CD25$^-$ CD4$^+$ (non Tregs) precursors are then cultured with SPE-K/L (or another superantigen) for 3 days. After this in vitro enrichment step, CD25$^+$ Tregs are isolated from the cell cultures and the Treg phenotype validated by Flow Cytometry. On the same day, in order to assess the Treg suppressive function on the proliferation of conventional T cell targets, a 4-day coculture assay is set up by mixing purified Tregs and conventional T cells in different ratios in the presence of feeder cells. On day 8, the proliferation of targets is measured by Flow Cytometry. Our results support that specific molecules derived from pathogens can directly modulate the differentiation of T cells into Tregs [17]. As Tregs are emerging as therapeutic targets for autoimmune [20] and cancer [21] disease, tools enabling their expansion, characterization, and manipulation will help future clinical developments.

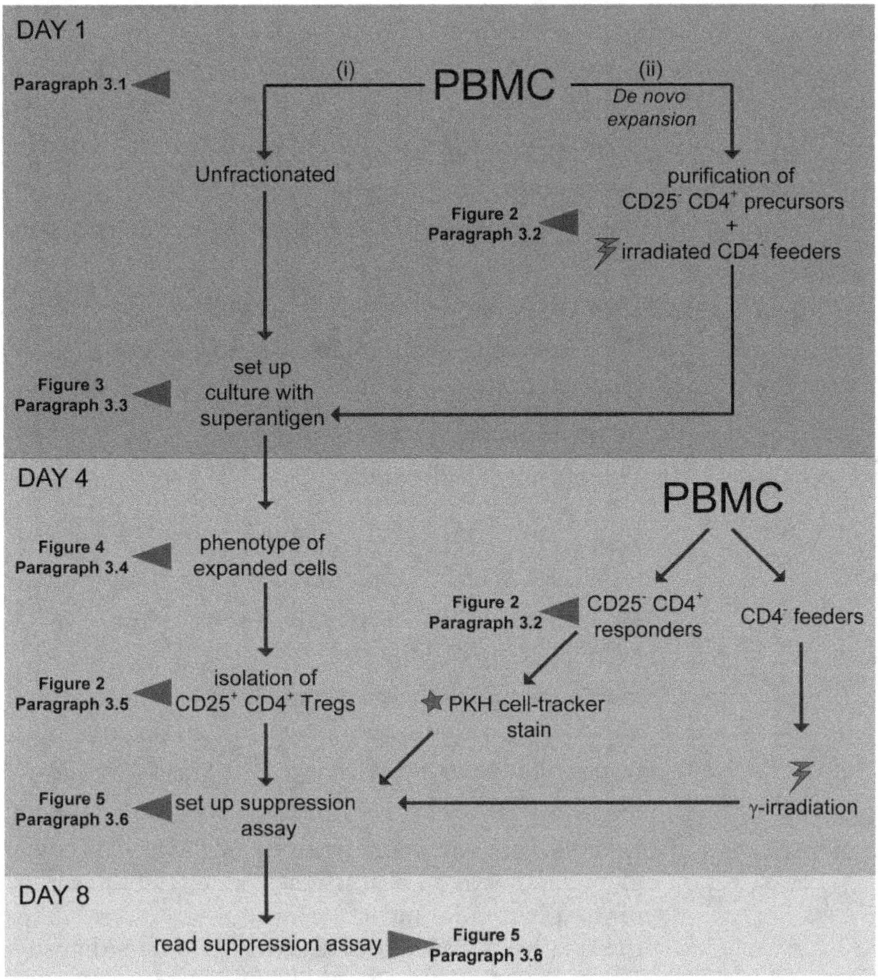

Fig. 1 Schematic overview of the experimental protocol. Using this protocol it is possible to: enrich Tregs with superantigen directly from unfractionated peripheral blood mononuclear cells (PBMCs), which does not exclude expansion from endogenous Tregs (*i*, refer to Subheading 3.1 and then Subheading 3.3.1); or isolate $CD25^-$ $CD4^+$ Treg-precursors from PBMC for the de novo superantigen-driven induction and expansion of Tregs (*ii*, refer to Subheading 3.1 then Subheadings 3.2 and 3.3.2). *DAY1.i.* In the direct expansion protocol, PBMCs are isolated (Subheading 3.1) and unfractionated cells seeded in culture with superantigen (Fig. 3 and Subheading 3.3.1). *DAY1.ii.* In the de novo induction protocol, after PBMC preparations (Subheading 3.1), $CD25^-$ $CD4^+$ (non Tregs) precursors are purified using a two-step separation protocol (Fig. 2 and Subheading 3.2) and the isolated precursors are seeded in culture with superantigen in the presence of autologous irradiated $CD4^-$ feeder cells (Fig. 3 and Subheading 3.3.2). The feeder cells need to be purified from the same donor in parallel to $CD25^-$ $CD4^+$ precursors (Subheading 3.2). *DAY4.* After 3 days of culture, $CD25^+$ Tregs are isolated from the cell cultures using a two-step separation protocol (Fig. 2 and Subheading 3.5) and the Treg phenotype is validated by Flow Cytometry (Fig. 4 and Subheading 3.4). On the same day, in order to assess the Treg suppressive function on the proliferation of conventional T cell targets, a coculture assay (Subheading 3.6) is set up mixing purified, in vitro superantigen-induced Tregs (Subheading 3.5) with freshly purified $CD25^-$ $CD4^+$ responders labeled with the proliferation cell tracker, PKH67, and irradiated feeder cells (both derived from PBMCs of the same donor as on day 1) in the presence of the stimulatory anti-CD3 antibody (OKT3). *DAY8.* The proliferation of targets is measured by Flow Cytometry (Fig. 5 and Subheading 3.6)

2 Materials

2.1 Venepuncture

1. Nitrile gloves.
2. Sodium-Heparin Vacutainers 170 I.U. (BD Biosciences).
3. Tourniquet.
4. Butterfly needles (BD Vacutainer Safety-Lok; Blood collection set).
5. Disposable plastic hubs (BD Vacutainer; One use holders).
6. 70 % isopropyl alcohol skin cleansing swabs.
7. Cotton swabs.
8. Sterile plasters.
9. Phlebotomy chair/room.

2.2 PBMC Preparation

1. Dulbecco's phosphate buffered saline (DPBS), pH 7.4. Stored at 4 °C once opened. Bring to room temperature (RT) before use.
2. Ficoll-Paque™ PLUS (GE Healthcare). Stored at RT. Use and keep in sterile conditions.
3. Sterile disposable *pasteurs.*
4. Trypan blue 0.4 % or 3 % acetic acid for viability counts. Dilute trypan blue to 0.1 % (1:4) in DPBS pH 7.4 and store at RT (keep sterile).
5. Roswell Park Memorial Institute (RPMI) 1640 medium: supplemented with 10 % heat-inactivated fetal calf serum (FCS), 2 mM L-glutamine, and 1 % penicillin–streptomycin. Referred in the text as cRPMI (complete RPMI—*see* Subheading 2.3.1). Filter cRPMI through a 0.22 μm filter to remove contaminants, crystals/precipitates that may adversely affect cell viability (keep sterile).
6. Stericup filter (Millipore units or similar).
7. Zero brake centrifuge.

2.3 Human Immune Cell Culture

2.3.1 General Tissue Culture

1. RPMI 1640 medium.
2. Fetal Calf Serum (FCS): heat deactivated for 1 h at 56 °C prior to use; stored in −20 °C aliquots.
3. L-glutamine: stored in −20 °C working aliquots.
4. Penicillin–streptomycin: stored in −20 °C aliquots.
5. SPE K/L (purified from *Escherichia coli* transfected with a spe-- K/L-expressing vector, a gift from Prof Thomas Proft—University of Auckland, New Zealand—as previously described [17, 22]).
6. SPEA (Toxin Technology).
7. TSST-1 (Sigma-Aldrich).

8. SEA (Toxin Technology).
9. OKT3 (eBioscience).
10. 24-wells plates.
11. 5 % CO_2 Incubator for mammalian cells.
12. CL-2 Safety Cabinet.
13. Pharmacidal or similar for mycoplasma containment in TC. Stored at room temperature and used as by manufacturer instructions.
14. Aquaguard-1 for disinfection of the water in TC incubator diluted as by manufacturer instructions. Stored in −20 °C working aliquots.
15. Aquaguard-2 for disinfection of the water in TC water baths diluted as by manufacturer instructions. Stored in −20 °C working aliquots.

2.3.2 Cell Isolation and Feeder Cell Irradiation

1. Human $CD4^+$ $CD25^+$ Regulatory T Cell Isolation Kit (Miltenyi Biotec). Stored at 2–8 °C, protected from light.
2. MACS® LD Column (Miltenyi Biotec).
3. MACS® MS Column (Miltenyi Biotec).
4. MACS® Separator (Miltenyi Biotec).
5. Bovine serum albumin (BSA).
6. 1 M EDTA for cell culture.
7. MACS buffer (homemade): DPBS, 0.5 % BSA, and 2 mM EDTA; filter through a 0.22 μm filter in sterile CL-2 cabinet to remove contaminants, crystals and precipitates that may adversely affect cell viability during the purification steps.
8. Stericup filter units (Millipore or similar).
9. 40 μm nylon mesh (BD Biosciences or similar).
10. Gamma-irradiator.

2.4 Flow Cytometry Reagents

1. Human FoxP3 Buffer Set (BD Pharmingen); stored at 4 °C, undiluted. Prepare working solutions of the BD Human FoxP3 Buffer Set as by the manufacturer instructions.
2. FACS buffer (staining buffer): 1× DPBS, 2 % heat-deactivated FCS, and 0.2 % NaN_3.
3. FACS tubes (5 mL tubes).
4. Anti-human CD4 FITC (BD Biosciences).
5. Anti-human TCR Vβ1 PE (Beckman Coulter).
6. Anti-human CD8 PE/Cy7 (BD Biosciences).
7. Anti-human CD25 ACP/H7 (BD Biosciences).
8. Anti-human CD3 Pacific Blue (BioLegend).
9. Aqua Live-Dead stain (Thermo Fisher Scientific).

10. Anti-human FoxP3 Alexa Fluor® 647 or APC (BD Biosciences).
11. Anti-human CD127 eFluor450 (eBioscience, Affymetrix).
12. Anti-human TCR Vβ1 FITC (Beckman Coulter).
13. Anti-human CTLA-4 PE (eBioscience, Affymetrix).
14. Anti-human CD4 ECD (Beckman Coulter).
15. Anti-human GITR PE/Cy5 (BioLegend).
16. Anti-human CD45RA QDot655 (Thermo Fisher Scientific).
17. BD FACS Lysing Solution (BD Biosciences): store at 4 °C, to be diluted in dH_2O (1:10) just before use.
18. PermBD2 (BD Biosciences): store at 4 °C, to be diluted in dH_2O (1:10) just before use.
19. PKH67 Fluorescent Cell Linker Kit (Sigma) made of PKH67 Dye (Sigma) and Diluent C (Sigma).
20. 37 % formaldehyde.
21. 10× DPBS.
22. 0.5 % formaldehyde: a freshly done 1:8 dilution of 4 % formaldehyde stock in FACS buffer. To make 4 % formaldehyde (100 ml): 75.2 ml dH_2O, 14 ml 10× DPBS, and 10.8 ml 37 % formaldehyde. Mix well and filter the solution through a 0.22 μm filter to remove precipitates/crystals in a CL-2 cabinet. Store sterile at 4 °C.

3 Methods

The experimental outline of this protocol is provided in Fig. 1.

3.1 Venepuncture, Blood Withdrawal, and PBMC Isolation

3.1.1 Blood Withdrawal from Healthy Donors by Venepuncture (Day 1)

1. Ethical approval must be in place for the donation of blood (*see* **Note 1**).
2. Label the Sodium-Heparin Vacutainers with the donor code and date.
3. Wash hands and wear nitrile gloves.
4. Place a tourniquet around the top of the donor's arm tightly, ensuring that a finger can pass underneath.
5. Assess the participant's arm for veins; typically the median cubital vein lying anterior to the elbow is punctured. Once the vein has been identified, disinfect the area with a 70 % isopropyl alcohol skin cleansing swab.
6. Attach the butterfly needle (*see* **Note 2**) to the appropriate plastic hub.
7. Insert needle into the vein at a ≤30° angle (*see* **Note 3**); if needle is correctly in vein, blood will flow into the loop attached to the needle.

8. Push the Sodium-Heparin Vacutainer tube with the lid into the plastic hub: the blood will start flowing from the loop into the tube automatically. Once the tube is full, remove tube from plastic hub, gently mix by reverting and repeat with other Vacutainer tubes if necessary, until the required amount of blood is withdrawn (*see* **Note 4**).
9. When the last Vacutainer is full, remove the tourniquet from the donor's arm.
10. Remove the butterfly needle at the same angle as when inserted and dispose of in the sharps bin.
11. Immediately after the needle is removed, press firmly on the skin with a cotton swab to stop the bleeding and apply a sterile plaster.
12. Dispose of gloves into clinical waste bin and wash hands.

3.1.2 PBMC Isolation from Healthy Human Volunteers by Density Centrifugation (Day 1)

1. Work in a CL-2 Safety Cabinet. Carefully open Sodium-Heparin Vacutainer tubes.
2. Transfer the heparinized blood to a new tube and dilute it 1:1 with sterile DPBS pH 7.4.
3. Carefully layer 35 mL of diluted blood over 15 mL of Ficoll-Paque™ PLUS in a 50 mL tube without disturbing the Ficoll-Paque™ PLUS. Handle the tube carefully not to disturb the set phases.
4. Centrifuge at 1000 × *g* for 20 min at RT with no brake, to separate PBMCs.
5. Handle the tube carefully not to disturb the set gradient. The layer of PBMC is the white opaque ring located at the interface between the Ficoll-Paque™ PLUS (bottom liquid fraction) and the plasma (top liquid fraction, yellow). Transfer the PBMCs to a fresh tube containing 10 mL of DPBS pH 7.4 with a disposable sterile *Pasteur* pipette to dilute off residues of Ficoll-Paque™ PLUS as soon as possible. The collection volume from each tube should be about 8–10 mL.
6. Fill up the collection tube with approximately 30 mL DPBS pH 7.4 to wash the cells and centrifuge at 300 × *g* for 10 min at RT.
7. Decant the supernatant, loosen the pellet, and resuspend it in 50 mL DPBS pH 7.4.
8. Centrifuge at 200 × *g* for 10 min at RT to remove platelets.
9. Decant the supernatant and resuspend the pellet in 2–10 mL DPBS pH 7.4 or cRPMI as required.
10. Count cells diluting a small test volume in 0.1 % trypan blue (1:1 suggested dilution) or in 3 % acetic acid (9:1 suggested dilution).
11. Adjust the volume with cRPMI to bring cells to a concentration of 10^6 cells/ml.

It is possible to enrich Tregs in cultures with superantigen directly from unfractionated PBMC (Fig. 1.i, refer to Subheading 3.3) or following isolation of $CD25^-$ $CD4^+$ Treg-precursors from PBMC in the presence of irradiated $CD4^-$ feeder cells (Fig. 1.ii, proceed to Subheading 3.2).

3.2 Purifications of CD4⁻ Feeder Cells and CD25⁻ CD4⁺ Treg-Precursors from Fresh PBMC

In Fig. 2, we highlight the column principle used in our protocol and give an outline of the series of purifications that lead to a set of cell-populations used in different parts of the protocol. As illustrated in Fig. 1, purifications may be performed directly from PBMCs at day 1 and day 4 or from stimulated cultured cells at day 4 to lead to the fractionation of (Fig. 2a):

1. $CD4^-$ feeder cells from fresh PBMCs (Fig. 2b).
2. $CD25^-$ $CD4^+$ precursors from fresh PBMCs (Fig. 2b, c).
3. $CD25^+$ $CD4^+$ T regs from cultures (Fig. 2b, c and *see* **Note 5**).

3.2.1 Preparation and Irradiation of CD4⁻ Feeder Cells (Day 1.ii and Day 4, Fig. 1) Using MACS Human CD4⁺ CD25⁺ Regulatory T Cell Isolation Kit

1. Work in a CL-2 Safety Cabinet.
2. Prepare PBMCs from anti-coagulated blood (refer to Subheading 3.1).
3. Centrifuge at $200 \times g$ for 10 min at RT to remove platelets.
4. Remove supernatant without perturbing the cell pellet and resuspend in cold DPBS. As suggested in the manufacturer instructions (Miltenyi Biotec), magnetic bead purifications usually require the cells to be kept cold at all times (usually on ice) and using precooled solutions when resuspending cells (*see* **Note 6**).

Fig. 2 (continued) Superantigen-induced $CD25^+$ $CD4^+$ Tregs are derived from cultures at day 4. (**b**) PBMCs or stimulated cultures are incubated with a cocktail of biotinylated antibodies directed against surface markers of non-$CD4^+$ cells and then anti-biotin magnetic beads. These suspensions are loaded onto a magnetic column in the presence of a magnetic field. Unlabeled $CD4^+$T cells do not bind and, hence, flow through the column despite the magnetic field (*left panel*). In contrast, magnetically labeled $CD4^-$ cells are retained within the column and are eluted by applying pressure after removing the column away from the magnetic field (*right panel*). The eluted $CD4^-$ cells include B cells, monocytes, DCs, and other MHC^+ cell-types, forming the bulk of antigen presenting cells (APC) useful as feeders after irradiation. (**c**) Magnetically unlabeled $CD4^+$ T cells (**b**, *left*) are incubated with anti-CD25 magnetic microbeads that bind cells expressing high levels of CD25 (a surface Treg marker). Cells are then loaded onto a new magnetic column in the presence of a magnetic field. Unlabeled $CD25^{-/low}$ ($CD4^+$ T) cells flow through the column despite the magnetic field (**c**, *left panel*). In contrast, magnetically labeled $CD25^+$ ($CD4^+$ T) cells are retained in the column and are eluted by applying pressure after removing the column away from the magnetic field (**c**, *right panel*). If the process starts with day 4 cultures, the eluted $CD25^+$ $CD4^+$ T cells are superantigen-iTregs, whilst the unlabeled $CD25^{-/low}$ $CD4^+$ T cells are cells overgrown in the cultures to be discarded. If PBMCs are used, the unlabeled $CD25^{-/low}$ $CD4^+$ T cells are the Treg precursors (*day 1.ii*) or the conventional targets of the suppression assay (*day 4*), whilst eluted $CD25^+$ $CD4^+$ T cells are endogenous Tregs (*see* **Note 5**). We recommend verifying the purity of recovered fractions (*refer to* Subheading 3.4.2 and **Note 16**)

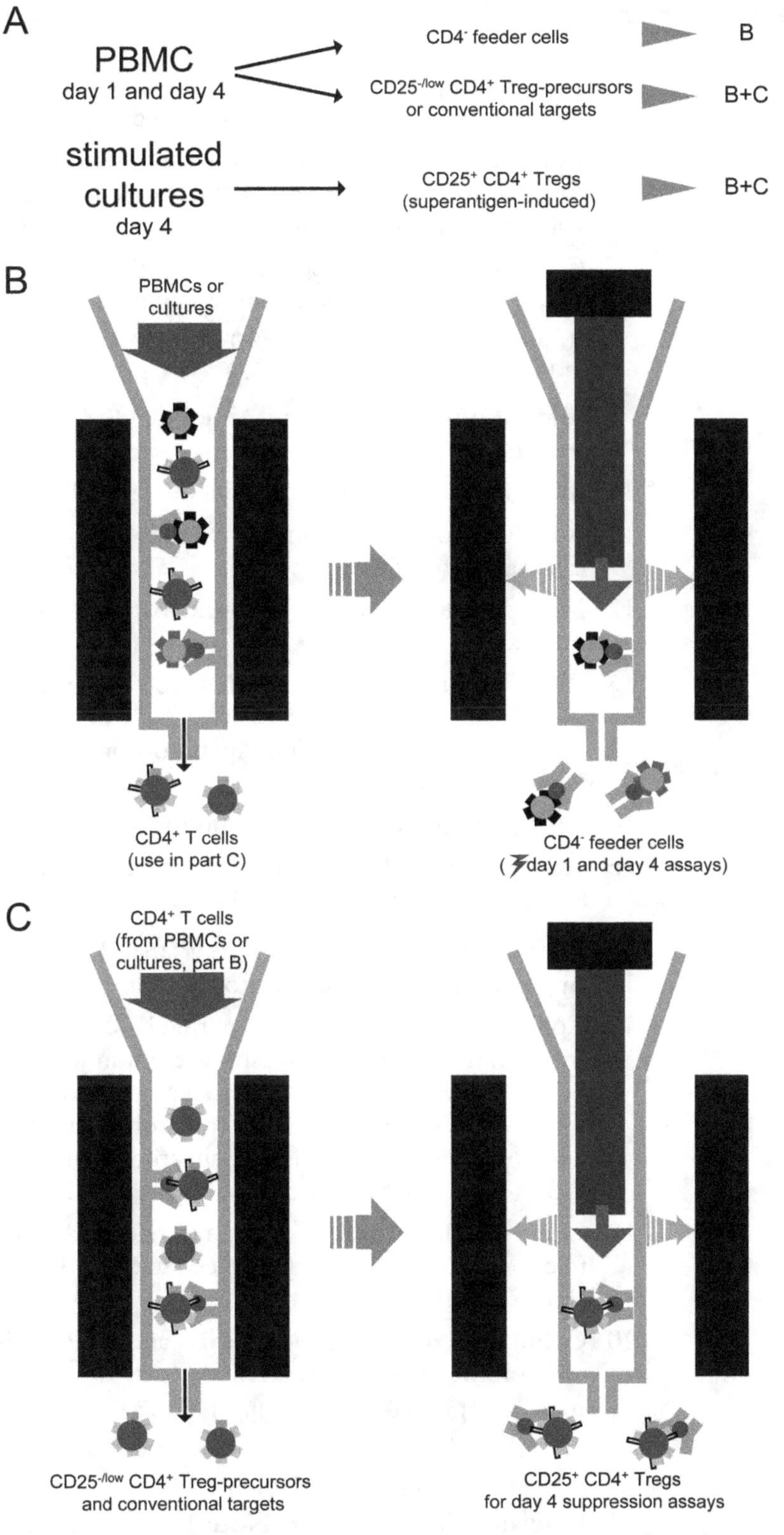

Fig. 2 Stepwise purification of cell-populations. (**a**) CD4⁻ feeder cells and CD25⁻ CD4⁺ precursors/conventional targets can be derived from fresh PBMCs: at day 1.ii for the de novo superantigen-driven induction and expansion of Tregs from CD25⁻ CD4⁺ precursors (Fig. 1 *DAY 1.ii*); and at day 4, for the suppression assay with CD25⁻ CD4⁺ conventional targets labeled with PKH67 (Fig. 1, *DAY 4*). CD4⁻ feeders are irradiated to avoid background proliferation.

5. For optimal performance (*see* **Note 7**) pass cells through a 40 μm nylon mesh to obtain a single cell suspension. Keep cells on ice.
6. Count cells diluting a small test volume in 0.1 % trypan blue. Any volume for magnetic labeling is based on 10^7 cells; scale up volumes according to the available cell number (*see* **Note 8**).
7. Centrifuge the cell suspension at 300 × *g* for 10 min. Pipette off or aspirate the supernatant completely.
8. Resuspend the cell pellet in 90 μL of MACS buffer *per* 10^7 cells (111×10^6 cells/ml).
9. Add 10 μL of Biotinylated-Antibody Cocktail *per* 10^7 cells to label non-$CD4^+$ cells.
10. Mix well (do not vortex) and incubate for 10 min on ice.
11. Add 20 μL of Anti-Biotin MicroBeads *per* 10^7 cells to magnetically tag non-$CD4^+$ cells.
12. Mix well (do not vortex) and incubate for additional 15 min on ice.
13. Wash the magnetically labeled cell suspension by adding 10–20× labeling volume of MACS buffer and centrifuge at 300 × *g* for 10 min at 4 °C. Pipette off or aspirate the supernatant completely.
14. Resuspend up to 10^8 cells in 500 μL of MACS buffer.
15. Place the LD Column in the magnetic field of a suitable MACS Separator placed inside the CL-2 cabinet.
16. Rinse the column with 2 mL of MACS buffer.
17. Apply the cell suspension onto the column.
18. Collect the unlabeled cells (purified $CD4^+$ cells) which pass through the column and wash the column with 1 mL of MACS buffer twice. Perform washing steps by adding MACS buffer successively once the column reservoir is empty. Collect total effluent cell fraction. This contains the unlabeled pre-enriched $CD4^+$ T cell fraction used in Subheading 3.2.2 (Fig. 2b, left).
19. Insert the plunger in the column to release the biotinylated cell fraction. These are the positively selected $CD4^-$ feeder cells (Fig. 2b, right).
20. Count cells recovered in **steps 18** and **19** by diluting a small test volume in 0.1 % trypan blue (*see* **Note 9**).
21. Resuspend $CD4^-$ feeder cells at 1×10^6/ml in cRPMI, secure the tube lids with Parafilm and proceed to γ-irradiation of the tube containing the cell suspension.
22. Irradiate cells (50 Gy) (*see* **Note 10** on alternatives).
23. Keep aside a small aliquot of irradiated $CD4^-$ feeder cells and unlabeled purified $CD4^+$ T cell to check the purity using Flow Cytometry (*see* Subheading 3.4.2).

3.2.2 Purifications of CD25⁻ CD4⁺ Treg-Precursors Prior to Treg Enrichment Culture from Unlabeled Purified CD4⁺ T Cell Fractions (Day 1.ii and Day 4; Fig. 1)

This protocol follows the purification of $CD4^{+}$ cells described in Subheading 3.2.1

1. Working in a CL-2 Safety Cabinet, perform **steps 2–18**, **20** and **23** as described in Subheading 3.2.1 to obtain unlabeled purified $CD4^{+}$ T cells from PBMCs. These will contain both $CD25^{+}$ (endogenous Tregs) and $CD25^{-}$ (precursors) $CD4^{+}$ T cells.
2. Centrifuge cells at $300 \times g$ for 10 min. Pipette off or aspirate the supernatant completely.
3. Resuspend the cell pellet in 80 μL of buffer *per* 10^{7} cells (125×10^{6} cells/mL).
4. Add 20 μL of CD25 MicroBeads *per* 10^{7} cells that will tag directly $CD25^{+}$ cells.
5. Mix well and incubate for 15 min on ice.
6. Wash cells by adding 10–20× labeling volume of MACS buffer and centrifuge at $300 \times g$ for 10 min (at 4 °C). Pipette off the supernatant completely.
7. Resuspend up to 10^{8} cells in 500 μL of MACS buffer.
8. Place the MS Column in the magnetic field of a suitable MACS Separator.
9. Prepare the column by rinsing with 500 μL of MACS buffer.
10. Apply the cell suspension onto the column.
11. Rinse the tube with another 500 μL MACS buffer and add any residual cells to the column.
12. Collect the unlabeled cells ($CD25^{-/low}$) that pass through and wash column with 1 mL of MACS buffer twice. Perform washing steps by adding MACS buffer once the column reservoir is empty (Fig. 2c, left).
13. Pipette 2 mL of MACS buffer onto the column, immediately flush out the fraction containing magnetically labeled cells ($CD25^{+}$ $CD4^{+}$ T cells) by firmly applying the plunger supplied with the column. Magnetic separation: Positive selection of $CD4^{+}CD25^{+}$ Tregs (Fig. 2c, right).
14. Keep aside a small aliquot of $CD25^{-/low}$ and $CD25^{+}$ cells to check the purity and CD4 expression by Flow Cytometry (*see* Subheading 3.4.2).

3.3 Superantigen Stimulation Assay

T cells are incubated with SPE-K/L superantigen for 3 days to generate inducible Tregs directly from unfractionated PBMC (Fig. 1.i, Subheading 3.3.1) or from $CD25^{-}$ $CD4^{+}$ Treg-precursors in the presence of irradiated $CD4^{-}$ feeder cells (Fig. 1.ii, Subheading 3.3.2). This will generate $CD25^{+}$ $FoxP3^{+}$ $V\beta1^{+}$ Tregs in the cultures (Fig. 3). The same protocol could be used to induce Treg using other superantigens, such as SEA, SPEA and TSST-1 [17, 19] (*see* **Note 11**).

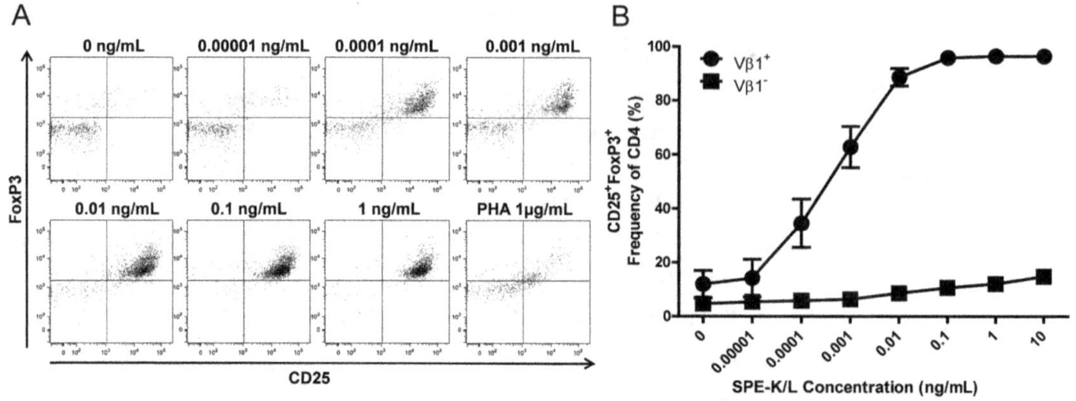

Fig. 3 Generation of $CD25^+$ $FoxP3^+$ $V\beta1^+$Tregs in superantigen-driven cultures. In order to validate the experimental conditions for this protocol, PBMCs were directly stimulated for 3 days with increasing concentrations of superantigen, SPE-K/L (ranging from 10^{-5} to 10 ng/mL) and, as controls, PHA (1 μg/ml) or cRPMI. After 3 days, cultured cells were analyzed by Flow Cytometry for TCR Vβ1, CD25, and FoxP3 expression (Subheading 3.4). (**a**) In a dose-dependent manner, the superantigen SPE-K/L induces the upregulation of CD25 and FoxP3 (the Treg signature transcription factor) in gated viable $CD3^+$ $CD4^+$ $V\beta1^+$T cells recovered from cultures. One of six donors is shown. (**b**) During the cultures, superantigen SPE-K/L preferentially induces TCR $V\beta1^+$, but not $V\beta1^-$ $CD4^+$ T cells, to express CD25 and FoxP3 in a dose-dependent manner. Events were gated for live $CD3^+$ $CD4^+$ T cells. Data represent the mean ± SEM of six donors. Figure reproduced with permission from the *Journal of Immunology* [17]. Copyright 2010. The American Association of Immunologists, Inc

3.3.1 In Vitro Culture in the Presence of Superantigen Directly from PBMC (Day 1.i through Day 4)

1. Seed PBMCs (10^6 cells) in 2 mL cRPMI *per* well of a 24-well plate. Seed multiple wells if necessary.
2. Add superantigen SPE-K/L (final concentration 1 ng/mL, Fig. 3 and *see* **Note 12**) to each well (other superantigens may be used alternatively, *see* **Note 11**).
3. Incubate the cell cultures for 3 days at 37 °C in a 5 % CO_2 incubator.
4. After 3 days in the presence of superantigens, the cells are harvested and replicate wells pooled (*see* Subheading 3.5).

3.3.2 In Vitro Culture in the Presence of Superantigen Directly from $CD25^-$ $CD4^+$ Precursors (Day 1.ii through Day 4)

1. Wash and incubate $CD25^{-/low}$ $CD4^+$ precursors (10^6 cells) together with irradiated $CD4^-$ feeder cells (10^6 cells) in a final volume of 2 mL of cRPMI *per* well of a 24-well plate.
2. Add to each well the superantigen SPE-K/L (final concentration 1 ng/mL, *see* **Note 12**); other superantigens may be used alternatively (*see* **Note 11**).
3. Incubate the cell cultures for 3 days at 37 °C in a 5 % CO_2 incubator.
4. After 3 days in the presence of superantigens, the cells are harvested and replicate wells pooled (*see* Subheading 3.5).

3.4 Treg Phenotype

In order to verify the presence and phenotype of Tregs we suggest using FoxP3 intracellular staining (Subheading 3.4.1) and/or a surface staining for the typical markers expressed by Tregs (Subheading 3.4.2) (Figs. 3 and 4). At day 1, PBMCs and any subfractionation product (as described in Subheadings 3.1.2 and 3.2) are tested for the phenotype of Tregs. At day 4, after harvesting the cultures (Subheading 3.3), an aliquot is analyzed by Flow Cytometry before proceeding with the purification of CD25$^+$ CD4$^+$ Tregs.

3.4.1 FoxP3 Intracellular Staining of CD4$^+$ T Cells (Day 1 and 4)

In the FoxP3 intracellular staining protocol, cells are resuspended and stained for surface markers (such as CD4 and CD25; **steps 1–9**), fixed and permeabilized (**steps 10–17**) and incubated with an antibody directed against anti-human FoxP3 for the detection of the Treg "signature" transcription factor (**steps 18–24**) (Fig. 3).

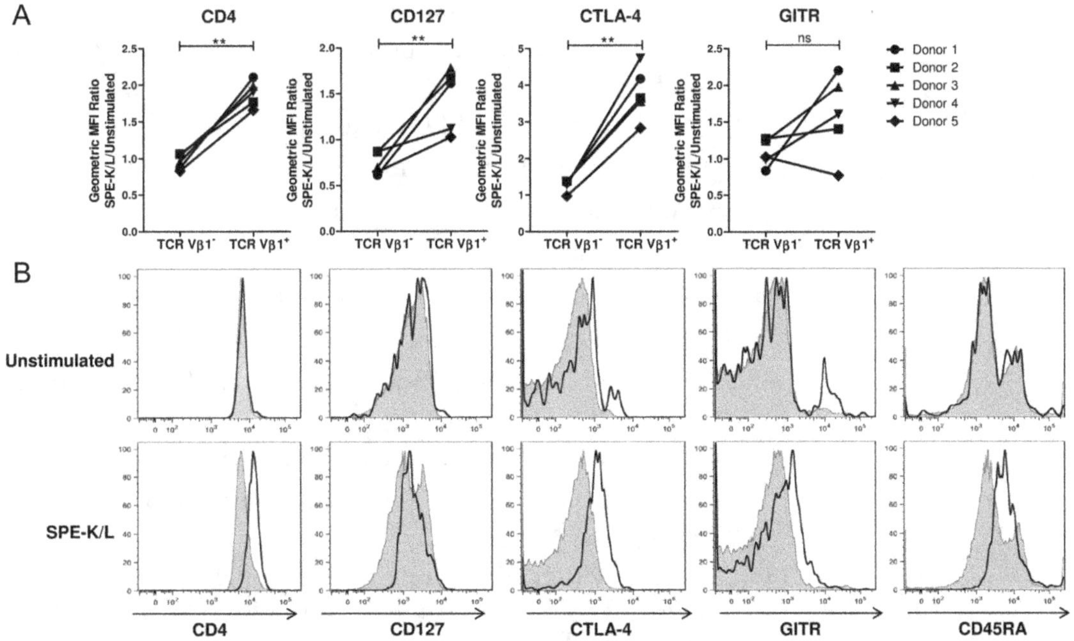

Fig. 4 Superantigen-induced TCRVβ1$^+$T cells express activation and regulatory T cell markers. PBMCs were cultured with SPE-K/L (1 ng/ml) or left unstimulated in cRPMI (control) for 3 days. On day 3, Flow cytometry was used to analyze the expression of a panel of activation and regulatory surface markers, including CD127, CTLA-4, GITR, and CD45RA in Vβ1$^+$ and Vβ1$^-$ CD4$^+$T cells (Subheading 3.4.2). (**a**) Following SPE-K/L stimulation, CD3$^+$ CD4$^+$Vβ1$^+$T cells (superantigen-iTreg) show increased expression levels of CD4, CD127, and CTLA-4 compared to CD3$^+$CD4$^+$Vβ1$^-$T cells, as expressed as geometric mean fluorescence intensity (MFI) ratio in SPE-K/L-stimulated compared to unstimulated cells, for any marker (five donors). (**b**) Vβ1$^+$ superantigen-iTregs express activation and regulatory T cell phenotype markers. Representative histograms show expression profile of CD4, CD127, CTLA-4, GITR, and CD45RA in gated, viable Vβ1$^-$ (*grey filled*) and Vβ1$^+$ (*black lines*) CD3$^+$ CD4$^+$T cells from control (unstimulated) and SPE-K/L-stimulated cultures. Figure reproduced with permission from the *Journal of Immunology* [17]. Copyright 2010. The American Association of Immunologists, Inc

1. Bring the Human FoxP3 Buffer Set (made of a Buffer A and a Buffer B) to RT.
2. Dilute the FoxP3 Buffer A (which is a 10× concentrate) with double distilled water at RT (1:10 ratio) to make 1× FoxP3 Buffer A.
3. FoxP3 Buffer C is then prepared by diluting FoxP3 Buffer B into 1× FoxP3 Buffer A (Buffer B:Buffer A ratio 1:50).
4. Prepare human PBMCs or superantigen-stimulated cultured cells (as specified above), count and resuspend the cells in FACS buffer at 10^7 cells/mL.
5. Dispense 100 μL of cell suspension into 5-mL tubes (for Flow Cytometry)
6. Prepare an antibody mix topping up volumes with staining buffer, so that the volume added to the each tube is 50 μL (*see* **Note 13**). Pipette appropriate amount of surface staining antibody to appropriate tubes. Create tubes for each single cell stains and the non-stained controls.
7. A typical surface marker antibody panel from our study (*see* **Note 14**):
   (a) Anti-human CD4 FITC; 5 μL/test
   (b) Anti-human TCR Vβ1 PE; 10 μL/test
   (c) Anti-human CD8 PeCy7; 1 μL/test
   (d) Anti-human CD25 ACP/H7; 5 μL/test
   (e) Aqua (live-dead stain); 0.75 μL/test
   (f) FACS buffer; 28.25 μL/test
8. Mix well and incubate cells with staining antibody for 20 min at 4 °C, in the dark (*see* **Note 15**).
9. Add 2 mL FACS buffer to each tube to dilute antibody.
10. Centrifuge at 300 ×*g* for 10 min (at RT) and remove the supernatant by decanting.
11. To fix the cells gently resuspend the cell pellets in the residual volume of FACS buffer left in the tubes after decanting. Add 2 mL of 1× FoxP3 Buffer A and vortex.
12. Incubate cells in fixative for 10 min at RT, in the dark (*see* **Note 15**).
13. Centrifuge at 500 ×*g* for 5 min. Remove fixative by pipetting (as pellet is buoyant).
14. Add 2 mL FACS buffer to the cells and centrifuge at 500 ×*g* for 5 min. Remove the wash buffer by decanting tubes.
15. To permeabilize the cells gently resuspend pellets in the residual volume of FACS buffer left in the tubes after decanting. Add 0.5 mL of 1× FoxP3 Buffer C to each tube and vortex.
16. Incubate cells with permeabilization reagent for 30 min at RT, in the dark (*see* **Note 15**).

17. Add 2 mL FACS buffer in each tube and centrifuge at 500 × *g* for 5 min at RT.
18. Decant supernatant, repeat **step 17** and decant supernatant again. Cells are now ready for the intracellular staining.
19. Prepare the following mix of antibodies and add 50 μL volume of mix to each tube:
    (a) Anti-human CD3 Pacific Blue (*see* **Note 25**); 1 μL/test
    (b) Fluorophore-conjugated (we suggest using Alexa Fluor® 647 or APC-labeled) anti-human FoxP3 antibody; 15 μL/test
    (c) FACS buffer; 34 μL/test
    Gently shake or vortex tubes to resuspend the pellet in each tube.
20. Incubate for 30 min at RT, in the dark (*see* **Note 15**).
21. Add 2 mL FACS buffer in each tube and centrifuge at 500 × *g* for 5 min at RT.
22. Decant supernatant, repeat **step 21,** and decant the supernatant again.
23. Resuspend in FACS buffer and analyze immediately.
24. (Optional) Add 300 μL of 0.5 % formaldehyde in each tube, mix and store at 4 °C, in the dark (*see* **Note 15**). Acquire cells within 24 h.
25. Acquire at least 15,000–25,000 $CD4^+$ lymphocytes on the Flow Cytometer (Fig. 3a).

3.4.2 Surface Marker Phenotype of Tregs and Purity Checks (Day 1 and 4)

Examining the surface marker phenotype of the purified Treg cells is suggested to confirm the presence of Tregs in the experiments (Fig. 4). The purity checks are necessary after each purification step using MACS columns. Irradiated (or if preferred not irradiated) $CD4^-$ feeder cells and unlabeled pre-enriched $CD4^+$ T cells as well as their sub-fractionated populations: $CD25^{-/low}$ and $CD25^+$ $CD4^+$ T cells all need to be checked by Flow Cytometry for the presence of CD4 and CD25 expression.

1. Prepare human PBMCs or cultured cells (as specified above), count and resuspend the cells in FACS buffer at 10^7 cells/mL.
2. Dispense 100 μL of cell suspension into 5-mL tubes (for Flow Cytometry)
3. Prepare a mix of the surface antibodies in FACS buffer, so that the final volume that is added to each tube is at least 50 μL (*see* **Note 13**). Create tubes for each single cell stain and the not-stained controls.
   For surface marker evaluation (Fig. 4) use the following antibodies (*see* **Note 14**):
    (a) Anti-human TCR Vβ1 FITC; 10 μL/test (*see* **Note 11**)
    (b) Anti-human CTLA-4 (CD152) PE; 20 μL/test

(c) Anti-human CD4 ECD; 3 μL/test
(d) Anti-human GITR PE/Cy5; 20 μL/test
(e) Anti-human CD8 PE/Cy7; 1 μL/test
(f) Anti-human CD25 APC/H7; 5 μL/test
(g) Anti-human CD45RA QDot655; 1 μL/test
(h) Aqua Live/Dead cell stain; 0.75 μL/test
(i) Anti-human CD127 eFluor450; 5 μL/test
(j) Add 65.75 μL of the mix to each tube

For purity check after fractionation of PBMCs (*see* **Note 16**), use the following antibodies:

(a) Anti-human CD4 ECD; 3 μl/test
(b) Anti-human CD8 PE/Cy7; 1 μl/test
(c) Anti-human CD25 APC/H7; 5 μl/test
(d) Anti-human CD3 Pacific Blue; 1 μl/test
(e) FACS buffer; 40 μl/test

Mix well and incubate cells with staining antibodies for 30 min at 4 °C, in the dark (*see* **Note 15**).

4. Add 2 mL FACS buffer to each tube to dilute antibody. Mix well.
5. Centrifuge at 300 × *g* for 10 min at RT
6. Remove supernatant by decanting and gently resuspend the cells in any residual volume of FACS buffer left in the tubes.
7. Add 1 mL of 1× BD FACS Lysing Solution (1:10 diluted in distilled water) to every tube, mix and incubate for 10 min, at RT, in the dark.
8. Add 3 mL of FACS buffer to each tube, mix and centrifuge (4 °C, 400 × *g*, 8 min). Decant the supernatant.
9. Keep samples at 4 °C and light protected until acquisition.
10. (Optional) Dispense 500 μL of 0.5 % formaldehyde in each tube, mix and store samples at 4 °C, in the dark (*see* **Note 15**). Acquire cells within 48 h (*see* **Note 17**).

3.5 Regulatory T Cell ($CD4^+$ $CD25^+$) Isolation from Superantigen-Driven Expansion Cultures

3.5.1 Harvesting 3-Day Cultures (Day 4)

1. Work in a CL-2 Safety Cabinet.
2. Assess the cultures at the microscope prior to proceeding to the cell harvest.
3. Replicate culture wells are harvested using a sterile mono-use *Pasteur*, pipetting multiple times around each well (avoid making bubbles) in order to ensure that each well contains minimal residual cells. The cell-suspension is then transferred to a 50 mL falcon collection tube.

4. To increase the yield of harvest, cold DPBS pH 7.4 may be added to each well after **step 3** and, after 5 min incubation, a second harvest is repeated as described in **step 3**. Pool any collected volume into the same collection tube.
5. Add more DPBS pH 7.4 to bring volume to 50 mL and centrifuge at 300×*g* for 10 min at RT.
6. Decant supernatant.
7. Resuspend pellet in 50 mL of cold DPBS pH 7.4 and centrifuge at 300×*g* for 10 min at 4 °C.
8. Carefully remove the supernatant and resuspend in 5 mL cold DPBS pH 7.4; keep cell suspensions in ice.
9. Test a small aliquot of the culture for the presence and phenotype of Tregs as described in Subheading 3.4.
10. Proceed to Subheading 3.5.2.

3.5.2 Purification of CD25⁺ CD4⁺ Tregs Expanded by Superantigen (Day 4)

1. Work in a CL-2 Safety Cabinet.
2. Repeat **steps 5–18** as described in Subheading 3.2.1 to eliminate any non-CD4 cells that may have been proliferating during the 3-day culture.
3. Count the negatively purified $CD4^+$ cells by diluting a small test volume in 0.1 % trypan blue.
4. Repeat **steps 1–13** as described in Subheading 3.2.2 to obtain positively enriched $CD25^+$ $CD4^+$ T cells.
5. Test a small aliquot (preferentially as in Subheading 3.4.1, *see* **Note 16**) for purity and phenotype.

3.6 Suppression Assay

In order to validate that Tregs are effective suppressor cells, the ability of $CD25^+$ $CD4^+$ T cells purified from the cultures with superantigen to suppress the proliferation of conventional responder, $CD25^-$ $CD4^+$ (or $CD8^+$) cells is measured in mixed cultures in the presence of irradiated $CD4^-$ feeder cells and stimulatory anti-CD3 antibody (OKT3), as described in this paragraph. Conventional responder ($CD25^-$ $CD4^+$) and irradiated feeder ($CD4^-$) cells are recovered from fresh blood, on day 4, from the same donor used to set up superantigen-driven cultures on day 1 (Fig. 5).

3.6.1 Purification of CD4⁻ Feeder Cells from the Same Donor Blood and Irradiation (Day 4)

On day 4, prepare $CD4^-$ feeder cells from fresh PBMCs (Fig. 2b) and irradiate them as previously detailed in Subheading 3.2.1, **steps 1–23**.

3.6.2 Purification of Responder, CD25⁻ CD4⁺ T Cells from the Same Donor Blood (Day 4)

On day 4 from fresh PBMCs, prepare $CD25^-$ $CD4^+$ cells starting from the unlabeled purified $CD4^+$ T cells obtained as the effluent fraction (as in Subheading 3.2.1, **step 18**; Fig. 2b, left) of the column used in Subheading 3.6.1. Using this fraction, proceed to

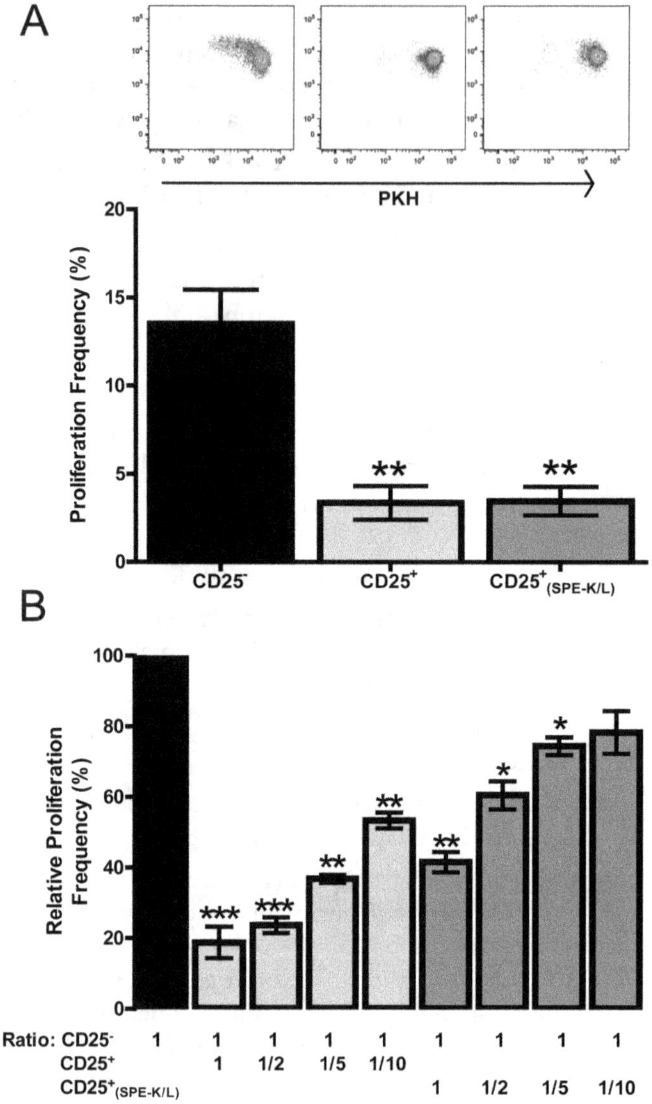

Fig. 5 SPE-K/L iTregs functionally suppress effector T cells. CD4$^+$ CD25$^+$ T cells were isolated from PBMCs stimulated with SPE-K/L (1 ng/ml) for 3 days as described in Subheadings 3.1, 3.3.1 and 3.5 (superantigen-iTregs). On day 4 (Fig. 1), irradiated CD4– feeder (Subheading 3.6.1) and PKH67-labeled CD4$^+$ CD25$^-$ T (Subheadings 3.6.2 and 3.6.3) cells were freshly derived from the same donor and cocultured with purified superantigen-iTregs in the presence of anti-CD3 for 4 days (Subheading 3.6.4) before Flow Cytometry analysis (Subheading 3.6.5). (**a**) SPE-K/L iTregs (CD25$^+$ SPE-K/L) suppress the proliferation of conventional CD4$^+$ CD25$^-$ T cells with the same strength as endogenous Tregs (CD25$^+$; freshly derived from PBMCs, *see* **Note 5**). Proliferation of conventional CD4$^+$ CD25$^-$ T cells was analyzed by the percentage of cells with decreased PKH67 expression. Data represent mean ± SEM of three donors. (**b**) Suppression by SPE-K/L iTregs (CD25$^+$SPE-K/L) depends on the number of iTregs seeded in the cultures relative to conventional T cell targets (ratio). Proliferation relative to maximal proliferation frequency is displayed. Data represent mean of triplicates from one donor ± SEM. Figure reproduced with permission from the *Journal of Immunology* [17]. Copyright 2010. The American Association of Immunologists, Inc

Subheading 3.2.2 (**steps 1–14**) to enrich $CD25^-$ (and $CD25^+$) $CD4^+$ T cells (Fig. 2c). $CD25^-$ $CD4^+$ T cells (Subheading 3.2.2, **step 12**) are then labeled with the proliferation cell tracker, PKH67 as in Subheading 3.6.3.

If desired, $CD25^+$ $CD4^+$ T cells (*i.e.*, endogenous Tregs) freshly recovered from the donor (in parallel to $CD25^-$ $CD4^+$ T cells Subheading 3.2.2, **step 13**) may be compared to $CD25^+$ $CD4^+$ T cells induced in superantigen cultures (*i.e.*, superantigen-induced Tregs, Subheading 3.5) for their suppressive strength (*see* **Note 18**). Our previous data suggest that superantigen-induced Tregs are as suppressive as endogenous Tregs (Fig. 5) [17].

3.6.3 PKH67 Staining of Responder $CD25^-$ $CD4^+$ T Cells (Day 4)

Immediately after purification from PBMCs at day 4, fresh $CD25^-$ $CD4^+$ T (conventional target) cells are labeled with a proliferation cell tracker (PKH67) according to the manufacturer instructions. During the PKH67 labeling phase, cells are diluted to 10^7 cells/mL and the working concentration of the cell-tracker is kept at 2×10^{-6} M (*see* **Note 19**).

These labeled cells will subsequently be cocultured with superantigen-induced Tregs, irradiated $CD4^-$ feeder cells and stimulatory anti-CD3 antibody. After 4 days of coculture, the frequency of proliferating ($PKH67^{low}$) cells is measured by Flow Cytometry to assess the suppression exerted by Tregs over the proliferation of the labeled T cell targets.

1. Work at RT.
2. Count $CD25^-$ $CD4^+$ cells eluted from the column at **step 12** of Subheading 3.2.2, diluting a small test volume in 0.1 % trypan blue.
3. Centrifuge $CD25^-$ $CD4^+$ cells at $300 \times g$ for 10 min. Decant the supernatant.
4. Add 15–50 ml DPBS pH 7.4 or RPMI medium (without serum or additives) to resuspend the cells and spin $300 \times g$ for 10 min.
5. Pipette off or aspirate the supernatant completely, leaving no more than 25 μL residue.
6. Prepare a 2× cell suspension by adding enough Diluent C to the cell pellet and resuspend the cells by pipetting gently, so that a homogenous cell suspension (2×10^7 cells/mL) is obtained. Do not vortex cells or incubate cells in Diluent C. Take note of the volume of Diluent C added.
7. Immediately prior to staining, in a fresh polypropylene falcon tube, prepare a 2× PKH67 Dye Working Solution (4 μM) in Diluent C by adding 4 μL of the PKH67 ethanolic dye solution *per* 1 mL of Diluent C, in enough Diluent C as required in **step 5** plus some excess volume. Vortex well the dispersion (60 s).

8. Rapidly add to the 2× cell suspension an equal volume of the 2× PKH67 Dye Working solution and immediately mix (crucial step) the sample by pipetting, so that during labeling cells are diluted to 10^7 cells/mL and the PKH67 dye to 2 μM final. Do not vortex (*see* **Note 20**).
9. Incubate the cell–dye suspension for 2 min with periodic mixing (*e.g.*, pipetting gently the cell suspension). Longer incubations will result in brighter cell staining and increased cell-death (*see* **Note 21**).
10. Stop the cell labeling by adding twice the volume as identified in **step 6** of FCS and incubate for 1 min to allow binding any dye excess.
11. Centrifuge the cells at 400×*g* for 10 min at RT and carefully remove the supernatant not to disturb pellet.
12. Resuspend the cell pellet in 10 mL of cRPMI and transfer to a fresh falcon tube.
13. Centrifuge at 400×*g* for 5 min at RT.
14. Repeat **steps 12** and **13** twice more to ensure removal of unbound dye.
15. After the final wash step, resuspend the cell pellet in 10 mL of cRPMI and dilute a small test volume in 0.1 % trypan blue for viability assessment.
16. Centrifuge at 400×*g* for 5 min and resuspend to the desired concentration of viable cells.

3.6.4 Seeding Conditions of Treg Suppression Assay and Culture (Day 4)

1. The $CD4^-$ feeder cells irradiated as discussed before (50 Gy at 10^6 cells/mL) are washed twice in cRPMI and resuspended in cRPMI at 10^6 cells/mL.
2. Set up the following incubations in a U-(round)-bottom plate in 200 μL volume, with technical triplicates. For the 1:1 (Tregs–conventional targets; Fig. 5) ratio (*see* **Note 22**), add 50,000 cells for each cell type (at 10^6 cell/mL; *i.e.*, 50 μL) in the following order:
   (a) $CD4^+$ $CD25^+$ Treg purified from cultures
   (b) $CD4^+$ $CD25^-$ responder cells labeled with PKH67 as in Subheading 3.6.3
   (c) Irradiated $CD4^-$ feeder cells
3. Add 50 μL OKT3 (working stock 2 μg/mL for a final concentration 0.5 μg/mL) and incubate for 4 days.

3.6.5 Reading the Treg Suppression Assay (Day 8)

1. Work in a CL-2 Safety Cabinet.
2. Assess the cultures at the microscope prior to proceeding to the cell harvest.
3. Culture wells are individually harvested by pipetting multiple times around each well (avoid making bubbles) in order to

ensure that each well contains minimal residual cells. Cell suspensions are then collected into 5 mL FACS tubes.

4. If it is necessary to increase the yield of harvest, 200 μL cold DPBS pH 7.4 may be added to each well after **step 3** and, after 5 min incubation, a second harvest is repeated as described in **step 3**. Pool any collected volume at the first and second harvests into the same collection tube.
5. Centrifuge tubes at 300 × *g* for 10 min at RT
6. Remove supernatant by decanting and gently resuspend the cells in any residual volume of FACS buffer left in the tubes.
7. Stain surface markers to detect proliferation using the following surface antibody mix (which takes into account PKH67 fluorescence; *see* **Note 23** and Fig. 5).
   (a) Aqua Live/Dead cell stain; 0.75 μL/test
   (b) Anti-human CD4 ECD; 3 μL/test
   (c) Anti-human CD8 PE/Cy7; 1 μL/test
   (d) FACS buffer; 45.25 μL/test

   As indicated above, prepare a mix of the surface antibodies in FACS buffer, so that the final volume that is added to each tube is 50 μL. Create tubes for each single cell stain and the non-stained controls (*see* **Note 24**).
8. Mix well and incubate cells with staining antibodies for 30 min at 4 °C, in the dark (*see* **Note 15**).
9. Add to any tube 2 mL FACS buffer to dilute antibody. Mix well.
10. Centrifuge at 300 × *g* for 10 min at RT
11. Remove supernatant by decanting and gently resuspend the cells in any residual volume of FACS buffer left in the tubes.
12. Add 1 mL of 1× BD FACS Lysing Solution (1:10 diluted in distilled water) to every tube, mix and incubate for 10 min, at RT, in the dark (*see* **Note 15**).
13. Add 3 mL of FACS buffer to each tube, mix and centrifuge (4 °C, 400 × *g*, 8 min). Decant the supernatant.
14. Add 1 mL of PermBD2 permeabilization solution (to obtain this, PermBD2 is diluted in distilled water 1:10) to every tube, mix and incubate for 10 min, at RT, in the dark (*see* **Note 15**).
15. Add 3 mL of FACS buffer to each tube, mix and centrifuge (4 °C, 400 × *g*, 8 min). Decant the supernatant.
16. After permeabilization, stain CD3 marker both intracellular and on the cell-surface (*see* **Note 25**):
   (a) Anti-human CD3 Pacific Blue; 1 μL/test
   (b) FACS buffer; 49 μL/test

Prepare a mix of the surface antibodies in FACS buffer, so that the final volume that is added to each tube is 50 μL. Create tubes for each single cell stain and the not-stained controls (*see* **Note 24**).

17. Mix well and incubate cells with staining antibodies for 30 min at 4 °C, in the dark (*see* **Note 15**).
18. Add 3 mL of FACS buffer to each tube, mix and centrifuge (4 °C, 400 × *g*, 8 min). Decant the supernatant.
19. Keep the samples at 4 °C and protected from the light until acquisition.
20. (Optional) Dispense 500 μL of 0.5 % formaldehyde in 1× PBS in each tube, mix and store the samples at 4 °C, in the dark (*see* **Note 15**). Acquire cells within 48 h (*see* **Note 17**).

4 Notes

1. We describe here the procedure that we routinely use in the lab to collect blood by venepuncture and the equipment typically needed; however, alternative protocols for venepuncture may be used to withdraw the blood. Ethical approval must be in place for the donation of blood and the donation must be performed by a trained phlebotomist after a patient information sheet has been discussed with the donor. The donor needs to complete and sign a consent form before the blood donation takes place. A record of the donations must be kept at the site and/or given to the donor (*e.g.*, as portable blood donor card) and any signed consent form must be kept in a locked area. A list of trained phlebotomists is also required to be kept at the site. We recommend Sodium-Heparin as anticoagulant when collecting blood for the preparation of PBMCs as used in this protocol.
2. Blue or green butterfly needles (BD) are used dependent on the blood donation volume for respectively ≤20 ml or between 20 and 50 ml of blood.
3. Anchor the vein by holding the participant's arm and placing a thumb below the venepuncture site. You can ask the participant to form a fist, so that the veins may appear more prominently—if needed. Ask the participant NOT to bend their arm; this increases the risk of bruising and hematoma.
4. Samples should be kept at RT until processing and left no longer than 6 h.
5. It is possible to purify the $CD25^+$ $CD4^+$Tregs from fresh PBMC at day 4 to compare them in parallel to superantigen-induced $CD25^+$ $CD4^+$ Tregs.

6. During the MACS cell separations, each step should be performed reasonably fast on ice or using cold solutions in sterile conditions.
7. This step is optional and—if desired—could be avoided. However, when putting cells through a magnetic column, we recommend filtering the cell suspensions through a nylon mesh of at least 70 μm (and ideally 40 μm) to avoid clamping the column. Similarly when running the samples in a flow cytometer (Subheading 3.4), filtering the samples through a nylon mesh may avoid clamping the fluidic parts of the machine.
8. Miltenyi Biotec recommends not scaling down volumes.
9. For any further use of CD4+ cells refer to Subheading 3.2.2.
10. As an alternative to irradiation, other protocols suggest using Mitomycin treatment to prepare feeder cells. If this method is preferred, please ensure that the cell viability in the cultures remains optimal and that the treatment does not compromise Treg expansion and function.
11. Different superantigens induce FoxP3^{+} Tregs of different TCR clonotype: we found SPE-K/L induces preferentially Vβ1^{+} Tregs whilst other superantigens that can also expand Tregs [17] elicit the expansion of different Vβ-repertoires [18, 19].
12. As shown in Fig. 3, we made a titration of superantigen to assess the optimal dose that maximally induces superantigen-iTregs. At 0.1 ng/mL SPE-K/L, 100 % of CD4^{+} Vβ1^{+} T cells were CD25^{+} FoxP3^{+} iTregs, also maintained at the higher concentrations of SPE-K/L tested (1–10 ng/ml). We then selected the mid-concentration (1 ng/ml) in the range inducing plateau-levels of iTregs. Refer to Refs. [17, 19] for further details and the titration data for the other superantigens.
13. Making a master mix containing all the antibodies useful in a Flow Cytometry panel (i) ensures that each sample gets exactly the same amount of each antibody and (ii) minimizes pipetting errors, so that any difference detected across the samples is reflective of biological differences in cell-marker expression rather than technical variability. We recommend preparing master mixes by topping up the summed volume of the different antibodies with staining buffer, up to 50 μl/sample (multiplying any amount by the number of tubes + 2 in excess). However, simply the sum of the mixed antibody volumes could be added to the tubes, preferred especially when the volume added is relatively large (>50 μl/sample).
14. We provide an example of Flow Cytometry master mix used routinely in the lab. First, the fluorophore combination was made compatible with our BD LSR II FACS machine settings;

after selecting specific antibody clones, antibody amounts were optimized by making titrations of each antibody—necessary to ensure that each stain is at saturation. We frequently—but not always—find that it is possible to reduce the amount of antibody down from the dose recommended by manufacturers. When designing a panel of antibodies different from the one proposed here, ensure that the fluorophore combination is compatible with the filter setting and instrumentation available at your institution. If the clone number (or the conjugated fluorophore) of any of the recommended antibodies is changed, then a titration of the new antibodies is needed prior to making any Flow Cytometry master mix.

15. Wrap the rack containing the tubes ready for Flow Cytometry acquisition in aluminum foil to avoid light exposure.
16. For the purity checks: if the frequency of $CD4^+$ cells in $CD25^{-/low}$ $CD4^+$ and $CD25^+$ $CD4^+$ sub-fractions is below 95 %, then the cells are to be loaded onto a second column prior to proceeding to any following step. Similarly if the frequency of $CD25^{hi}$ cells in the $CD25^{-/low}$ fraction is above 5 % (and conversely in the $CD25^+$ fraction below 95 %) then a second column purification is needed to fractionate cells according to CD25 expression. In the case of the purity checks after fractionation of superantigen-driven cultures at day 4, because of the downregulation of CD3 after stimulation in vitro (*see* **Note 25**), it may be preferable to use the protocol at Subheading 3.4.1 rather than the surface stain suggested in Subheading 3.4.2 (which however remains valid for the purity check of fractionations from fresh PBMCs at day 1 and day 4).
17. We recommend acquiring the Flow Cytometry samples on the day of sample processing, if possible. Ensure that your application is compatible with our guidelines for sample storage if you need to acquire them on a different day from the sample processing. In this case, samples stained with antibodies directed against surface antigens are more stable than samples permeabilized and stained with antibodies directed against intracellular antigens, even after fixation and at 4 °C. Hence we recommend not storing FoxP3 stained samples for more than 24 h, 4 °C. In contrast, samples stained for the detection of surface markers of for purity may be stored for 48 h or even over weekend at 4°C, if necessary.
18. During the culture with superantigen we have identified the presence of $CD25^+$ $CD8^+$Tregs [19]. These T cells belong to a different lineage from $CD4^+$ T cells and currently most reports focus on $CD4^+$ Tregs. However, we could demonstrate that bacterial superantigens can promote the induction of human $CD25^+$ $CD8^+$ Tregs, at least in vitro [19].

19. During the PKH67 labeling, some cells will die which is why we recommend counting cells before and after the procedure (*see* **step 2** and **15**). The optimal concentration of PKH67 enabling to detect proliferation and minimizing cell-death during the labeling protocol should be determined in the Lab. In our experience, optimal concentration will be found using PKH67 between 0.75 and 2 μM. If a final PKH67 concentration needed is less than 2 μM, the most reproducible results will be obtained by making an intermediate dye stock by diluting the PKH67 ethanolic dye solution provided in the kit with 100 % ethanol.
20. Rapid and homogeneous dispersion of the cells in solution is critical for uniform labeling, as the dye will attach to cells instantaneously. Use gentle pipetting to minimize cell-death.
21. Time during this step is crucially affecting both staining and cell-death during the labeling protocol. PKH67 labeling time should be adjusted in the specific Lab to balance intensity of labeling versus cell viability. We found that 2 min incubation is sufficient to stain the cells optimally, minimizing cell-death, and the manufacturer recommends incubation times between 2 and 5 min.
22. The Treg–conventional cell ratio may be titrated in cultures by changing the proportions of iTregs and responder T cells. In this case, ensure to keep the total number of cells in each well constant.
23. If it is needed to change fluorophores, exclude FITC, Alexa 488, and any similar dye with similar excitation and emission spectra, as this will interfere with detection of PKH67.
24. When working with PKH67 labeled cell cultures, it is recommendable to seed 1–2 wells in the same plate as described in Subheading 3.6.4, with $CD4^+$ $CD25^-$ responder cells not labeled with PKH67, by keeping a small aliquot of cells prior to performing the labeling described in Subheading 3.6.3 at day 4. These cells are the best non-stained control to use in Flow Cytometry on day 8. If it is decided to generate such control, we suggest stimulating $CD4^+$ $CD25^-$ responder cells not labeled with PKH67 as in Subheading 3.6.4, **step 3** (no feeder/Treg cells are needed for this purpose). It is recommendable to use this staining control to generate any other antibody stain control in Subheading 3.6.5. A PKH67 single stain will also be needed. This could be created by pooling a small fraction (5–10 μL) of any harvested condition just before **step 7** in Subheading 3.6.5.
25. We suggest staining for CD3 after permeabilization. This is necessary as CD3 is broadly downregulated in cultured cells in the presence of OKT3.

References

1. Sakaguchi S (2011) Regulatory T cells: history and perspective. Methods Mol Biol 707:3–17. doi:10.1007/978-1-61737-979-6_1
2. Schmitt EG, Williams CB (2013) Generation and function of induced regulatory T cells. Front Immunol 4:152. doi:10.3389/fimmu.2013.00152
3. Gratz IK, Campbell DJ (2014) Organ-specific and memory treg cells: specificity, development, function, and maintenance. Front Immunol 5:333. doi:10.3389/fimmu.2014.00333
4. Duhen T, Duhen R, Lanzavecchia A, Sallusto F, Campbell DJ (2012) Functionally distinct subsets of human FOXP3+ Treg cells that phenotypically mirror effector Th cells. Blood 119:4430–4440. doi:10.1182/blood-2011-11-392324
5. Schmetterer KG, Neunkirchner A, Pickl WF (2012) Naturally occurring regulatory T cells: markers, mechanisms, and manipulation. FASEB J 26:2253–2276. doi:10.1096/fj.11-193672
6. Zheng Y, Rudensky AY (2007) Foxp3 in control of the regulatory T cell lineage. Nat Immunol 8:457–462. doi:10.1038/ni1455
7. Lin W, Haribhai D, Relland LM, Truong N, Carlson MR, Williams CB, Chatila TA (2007) Regulatory T cell development in the absence of functional Foxp3. Nat Immunol 8:359–368. doi:10.1038/ni1445
8. Ohkura N, Hamaguchi M, Morikawa H, Sugimura K, Tanaka A, Ito Y, Osaki M, Tanaka Y, Yamashita R, Nakano N, Huehn J, Fehling HJ, Sparwasser T, Nakai K, Sakaguchi S (2012) T cell receptor stimulation-induced epigenetic changes and Foxp3 expression are independent and complementary events required for Treg cell development. Immunity 37:785–799. doi:10.1016/j.immuni.2012.09.010
9. Thornton AM, Korty PE, Tran DQ, Wohlfert EA, Murray PE, Belkaid Y, Shevach EM (2010) Expression of Helios, an Ikaros transcription factor family member, differentiates thymic-derived from peripherally induced Foxp3+ T regulatory cells. J Immunol 184:3433–3441. doi:10.4049/jimmunol.0904028
10. Weiss JM, Bilate AM, Gobert M, Ding Y, de Lafaille MA C, Parkhurst CN, Xiong H, Dolpady J, Frey AB, Ruocco MG, Yang Y, Floess S, Huehn J, Oh S, Li MO, Niec RE, Rudensky AY, Dustin ML, Littman DR, Lafaille JJ (2012) Neuropilin 1 is expressed on thymus-derived natural regulatory T cells, but not mucosa-generated induced Foxp3+ T reg cells. J Exp Med 209:1723–1742, S1721. doi:10.1084/jem.20120914
11. Yadav M, Louvet C, Davini D, Gardner JM, Martinez-Llordella M, Bailey-Bucktrout S, Anthony BA, Sverdrup FM, Head R, Kuster DJ, Ruminski P, Weiss D, Von Schack D, Bluestone JA (2012) Neuropilin-1 distinguishes natural and inducible regulatory T cells among regulatory T cell subsets in vivo. J Exp Med 209:1713–1722, S1711–1719. doi:10.1084/jem.20120822
12. Lin X, Chen M, Liu Y, Guo Z, He X, Brand D, Zheng SG (2013) Advances in distinguishing natural from induced Foxp3(+) regulatory T cells. Int J Clin Exp Pathol 6:116–123
13. Daniel C, Ploegh H, von Boehmer H (2011) Antigen-specific induction of regulatory T cells in vivo and in vitro. Methods Mol Biol 707:73–185. doi:10.1007/978-1-61737-979-6_11
14. Fantini MC, Dominitzki S, Rizzo A, Neurath MF, Becker C (2007) In vitro generation of CD4+ CD25+ regulatory cells from murine naive T cells. Nat Protoc 2:1789–1794. doi:10.1038/nprot.2007.258
15. Nouze C, Pasquet L, van Meerwijk JP (2011) In vitro expansion of alloantigen-specific regulatory T cells and their use in prevention of allograft rejection. Methods Mol Biol 707:187–196. doi:10.1007/978-1-61737-979-6_12
16. Chen Q, Kim YC, Laurence A, Punkosdy GA, Shevach EM (2011) IL-2 controls the stability of Foxp3 expression in TGF-beta-induced Foxp3+ T cells in vivo. J Immunol 186:6329–6337. doi:10.4049/jimmunol.1100061
17. Taylor AL, Llewelyn MJ (2010) Superantigen-induced proliferation of human CD4+CD25- T cells is followed by a switch to a functional regulatory phenotype. J Immunol 185:6591–6598. doi:10.4049/jimmunol.1002416
18. Llewelyn M, Sriskandan S, Terrazzini N, Cohen J, Altmann DM (2006) The TCR Vbeta signature of bacterial superantigens spreads with stimulus strength. Int Immunol 18:1433–1441. doi:10.1093/intimm/dxl076
19. Taylor AL, Cross EL, Llewelyn MJ (2012) Induction of contact-dependent CD8(+) regulatory T cells through stimulation with staphylococcal and streptococcal superantigens. Immunology 135:158–167. doi:10.1111/j.1365-2567.2011.03529.x
20. Fessler J, Felber A, Duftner C, Dejaco C (2013) Therapeutic potential of regulatory T cells in autoimmune disorders. BioDrugs 27:281–291. doi:10.1007/s40259-013-0026-5
21. Caserta S, Borger JG, Zamoyska R (2012) Central and effector memory CD4 and CD8 T-cell responses to tumor-associated antigens. Crit Rev Immunol 32:97–126
22. Proft T, Webb PD, Handley V, Fraser JD (2003) Two novel superantigens found in both group A and group C Streptococcus. Infect Immun 71:1361–1369

Chapter 17

Construction of Recombinant Single Chain Variable Fragment (ScFv) Antibody Against Superantigen for Immunodetection Using Antibody Phage Display Technology

Pawan Kumar Singh, Ranu Agrawal, D.V. Kamboj, and Lokendra Singh

Abstract

Superantigens are a class of antigens that bind to the major histocompatibility complex class (MHC) II and T-cell receptor (TCR) and cause the nonspecific activation of T cells, resulting in a massive release of pro-inflammatory mediators. They are produced by the gram-positive organisms *Staphylococcus aureus* and *Streptococcus pyogenes*, and by a variety of other microbes such as viruses and mycoplasma, and cause toxic shock syndrome (TSS) and even death in some cases. The immunodetection of superantigens is difficult due to the polyclonal activation of T-cells leading to nonspecific antibody production. The production of recombinant monoclonal antibodies against superantigens can solve this problem and are far better than polyclonal antibodies in terms of detection. Here, we describe the construction of recombinant single chain variable fragments (ScFv) antibodies against superantigens with specific reference to SEB (staphylococcal enterotoxin B) using antibody phage display technology.

Key words Superantigen, SEB, Recombinant antibody, ScFv, Phage display, Biopanning, Bacteriophages

1 Introduction

Bacterial superantigens (SAgs) are well described T cell mitogens and constitute a family of more than 40 secreted proteins and toxins. Unlike conventional peptide antigens, SAgs bypass normal processing by antigen presenting cells (APCs) and induce the activation of a large population (5–30 %) of T cells nonspecifically [1]. SAgs first bind, as intact molecules, to the MHC II on APCs outside of the peptide binding groove and then sequentially bind to the TCR via the variable region of the TCR Vβ chain [1–6]. This superficial binding of SAgs to the MHC-TCR complex leads to the activation of both APCs and T cells, resulting in a massive systemic release of pro-inflammatory cytokines such as TNF-α, IL-1β, and IFNγ and T cells mediators like IL-2, which can lead to fever and

Amanda J. Brosnahan (ed.), *Superantigens: Methods and Protocols*, Methods in Molecular Biology, vol. 1396,
DOI 10.1007/978-1-4939-3344-0_17, © Springer Science+Business Media New York 2016

toxic shock [1, 2, 5, 7]. The increased expression of adhesion molecules and changes in the chemokine gradient brought on by exposure to SAgs direct leukocyte migration to the sites of tissue injury [1, 8]. IL-2, expressed by SAgs activated T-cells, causes vasodilation, vascular leakage, and edema [9]. The ROS produced by the activated neutrophils increase vascular permeability and can cause acute lung injury [1, 8]. These molecular changes occur rapidly upon SAgs exposure and lead to hypotension, multi-organ failure, and even death [1, 6].

The SAg family consists of a number of protein toxins and toxin-like mitogens of about 20–30 kDa, produced by several species of gram-positive bacteria, including *Staphylococcus aureus*, *Streptococcus pyogenes*, and *Clostridium perfringens*, as well as the gram-negative bacteria *Yersinia pseudotuberculosis*, the wall-less *Mycoplasma arthritidis*, and even some viruses.

The detection of SAgs using polyclonal antibodies is difficult because the generation of polyclonal antibodies against SAgs is crippled by the fact that they are both low titered and nonspecific. A better alternative is the development of monoclonal antibodies against SAgs. In the past, hybridoma was the only method used in the production of monoclonal antibodies; however, this method is both laborious and time consuming. Also, many of the hybridoma clones producing monoclonal antibodies have been reported to be genetically unstable and tend to lose their antibody secreting ability with time [6, 10–12]. Recently, recombinant DNA technology and antibody engineering has improved the process of monoclonal antibody production and has made it possible to clone antibody genes in bacteria [13], mammalian cells and yeast [14], plant cells [15], and insect cells [16] using antibody phage display technology. Such 'immortalization' of antibody genes makes it technically feasible to produce recombinant monoclonal antibodies called single chain variable fragments (ScFv) quickly in bacterial cultures. These ScFv antibody molecules can further be genetically manipulated for improved specificity and affinity [17, 18]. The cost of their production is very low and they can be fused with marker molecules for the immunological detection of bacterial and viral agents, as well as other antigenic proteins [19–21]. Here, in this chapter, we describe the construction of recombinant ScFv antibodies against superantigens, specifically staphylococcal enterotoxin B (SEB), for immunodetection using antibody phage display technology.

2 Materials

RNase-free (DEPC treated) material (e.g., microtips, centrifuge tubes) should be used for all procedures involving RNA. Prepare all reagents using ultrapure nuclease-free water for the RNA isolation procedure. Clean and wipe the work bench, micropipettes, and any other instruments with RNaseAWAY® solution (Sigma-Aldrich) to destroy any RNase present. Use molecular biology grade reagents for all procedures.

2.1 Buffers and Reagents

1. 1× PBS: Dissolve 137 mM NaCl (8.0 g/l), 2.7 mM KCl (0.2 g/l), 10 mM Na_2HPO_4 (1.44 g/l), and 2 mM KH_2PO_4 (0.24 g/l) in 800 mL of deionized water. Adjust the pH to 7.4 and make up the volume to 1 l.
2. PBST: Add 0.05 % tween 20 (500 µl/l) to the 1× PBS.
3. MPBS: Dissolve the desired amount (2 g for 2 %, 5 g for 5 %, and 10 g for 10 %) of skim milk powder in 100 ml of 1× PBS.
4. Carbonate–Bicarbonate Buffer: Dissolve Na_2CO_3 (15.9 g/l), $NaHCO_3$ (29.3 g/l) in 800 ml deionized water, adjust the pH to 9.6, and make up the volume to 1 l.
5. PEG/NaCl Solution: Dissolve polyethylene glycol 8000 (200 g/l) and NaCl (146.1 g/l) in deionized water, dissolve by heating, and make up the volume to 1 l.
6. TES (1×): Dissolve 0.2 M Tris–HCl (24.2 g/l), 0.5 mM EDTA (0.15 g/l), 0.5 M sucrose (171.1 g/l) in 800 ml of deionized water, adjust the pH to 8.0 and make up the volume to 1 l.

2.2 Growth Media

1. 2× YT medium: Dissolve Bacto-tryptone (17.0 g/l), Bacto-yeast extract (10.0 g/l), and sodium chloride (5.0 g/l) in deionized water, make up the volume to 1 l, and sterilize by autoclaving.
2. 2× YT-G medium: Add 2 % (w/v) glucose to the 2× YT medium.
3. 2× YT-AG medium: Add 100 µg/ml ampicillin and 2 % (w/v) glucose to the 2× YT medium.
4. 2× YT AK medium: Add 100 µg/ml ampicillin and 50 µg/ml kanamycin to the 2× YT medium.
5. 2× YT-AI medium: Add 100 µg/ml ampicillin and 1 mM IPTG to the 2× YT medium.
6. SOB medium: Dissolve Bacto-tryptone (20.0 g/l), Bacto-yeast extract (5.0 g/l), and sodium chloride (0.5 g/l) in deionized water. Add 10 ml of 250 mM KCl and adjust the pH to 7.0. Make the volume up to 1 l and sterilize by autoclaving. Add 10 ml of 1 M $MgCl_2$ (filter sterile) prior to use.
7. SOB-AG medium: Add 100 µg/ml ampicillin and 20 g/l glucose to the SOB medium.
8. SOB-AGN medium: Add 100 µg/ml ampicillin 20 g/l glucose and 100 µg/ml nalidixic acid to the SOB medium.

3 Methods

3.1 Isolation of RNA from Hybridoma Cells or Immunized Mouse Spleen

1. Grow hybridoma cells (developed against the desired antigen) to around 80–90 % confluence for RNA isolation. [Alternatively, euthanize primed mice (mice immunized with the desired antigen) and isolate the spleen in aseptic condition] (*see* **Note 1**).
2. Discard the culture medium from the culture flask and wash the cells with sterile PBS.
3. Lyse the cells directly in the culture flask by adding 1 ml of TRIzol per 10 cm^2 of flask surface area.

   (For the mouse spleen, crush 1 mg of spleen tissue using a sterile, DEPC treated mortar and pestle in liquid nitrogen and resuspend the powdered spleen in 1–2 ml of TRIzol)
4. Homogenize the cell/tissue lysate by repeatedly passing it through a pipette or with a 21 gauge needle and allow it to stand for 5 min at room temperature for complete dissociation of the nucleoprotein complexes.
5. Add chloroform (0.2 ml/ml of TRIzol) to the cell/tissue lysate, shake vigorously for 15 s to create a uniform emulsion, and allow it to stand at room temperature for 5 min.
6. Centrifuge the cell/tissue lysates at 12,000 × *g* for 15 min at 4 °C to separate the organic phase (red in color, containing protein), an interphase (containing DNA), and a colorless aqueous phase (containing RNA).
7. Carefully transfer the upper aqueous phase to a new tube and add isopropanol (0.5 ml/ml of TRIzol) to precipitate the RNA (*see* **Note 2**).
8. Incubate the tube at room temperature for 15 min with occasional mixing by inverting the tube.
9. Centrifuge the samples at 12,000 × *g* for 10 min at 4 °C. Discard the supernatant and wash the RNA pellet with 1 ml of 75 % chilled ethanol per 1 ml of TRIzol, by centrifugation at 12,000 × *g* for 5 min at 4 °C.
10. Air-dry the RNA pellet (*see* **Note 3**) and dissolve the pellet in a suitable amount of nuclease-free water (30–50 μl). Make 2–3 aliquots, one each for quantitative (spectrophotometric) estimation, qualitative (denaturing agarose gel electrophoresis) estimation, and downstream processing, respectively. Store the remaining RNA at −20 °C for immediate use (*see* **Note 4**).
11. Estimate the quantity and quality (integrity) of the RNA using spectrophotometer/Nanodrop ($A_{260/280}$) and by running RNA on 1.2 % denaturing agarose gel, respectively.

3.2 Reverse Transcription for First Strand cDNA Synthesis

1. Reverse transcribe the total RNA (~1 μg) as described below using a primer mixture against the conserved 3′-end of antibody variable heavy (V_H) and variable light (V_L) region genes listed in Table 1. Prepare the reaction mixture as given below:

Reaction Mix (5×):	4.0 μl
Maxima Enzyme Mix:	2.0 μl
RNA:	1.0 μg
Primer (20 pmol/μl):	1.0 μl
Sterile nuclease-free water:	To make up the volume to 20.0 μl

2. Reverse-transcribe the above RNA mix using a thermal cycler using a cycle of 25 °C for 10 min, 50 °C for 40 min, and 85 °C for 5 min. Place on ice for further use.
3. Dilute the cDNA (1:10 dilution) and use 1 μl of diluted product as a template for the V_H and V_LPCR amplification steps.

3.3 PCR Amplification of Antibody Variable (V_H and V_L) Region Genes

1. Amplify the V_H and V_L regions of the immunoglobulin genes using hot start PCR. Use RS V_H For mix & Lin V_H Rev mix and Lin V_L For mix & RS V_L Rev mix degenerate primers for the amplification of the V_H and V_L chains, respectively (Table 1). These primers are designed in such a way that the V_H and V_L fragments will be synthesized with linker overhangs.
2. For 20 μl of reaction volume, prepare 12 μl of sample mix for each reaction, as described below. Heat the sample mix at 94 °C for 5 min and then add 8 μl of reaction mix and perform the PCR using the cycles given below.
3. Sample mix:

   Sample mix for V_H chain:

Template (1:10 dilution of cDNA):	1.0 μl
RS V_H For mix (20 pmol/μl):	1.0 μl
Lin V_H Rev mix (20 pmol/μl):	1.0 μl
Sterile nuclease-free water	to make up the volume to 12.0 μl

Sample mix for V_L chain:

Template (1:10 dilution of cDNA):	1.0 μl
Lin V_L For mix (20 pmol/μl):	1.0 μl
RS V_L Rev mix (20 pmol/μl):	1.0 μl
Sterile nuclease-free water	to make up the volume to 12.0 μl

Table 1
Primers used for amplification of variable heavy (V_H), variable light (V_L), and ScFv antibody genes (reproduced from Ref. [6])

Primer name	Sequence (5′ ′3′)
VHreverse	*Equimolar mixture of primers VHR 1, VHR 2, VHR 3, and VHR 4*
V_HR 1	GAG GAA ACG GTG ACC GTG GT
V_HR 2	GAG GAG ACT GTG AGA GTG GT
V_HR 3	GCA GAG ACA GTG ACC AGA GT
V_HR 4	GAG GAG ACG GTG ACT GAG GT
VLreverse	*Equimolar mixture of primers VLR 1, VLR 2, VLR 3, and VLR 4*
V_LR 1	ACG TTT KAT TTC CAG CTT GG
V_LR 2	ACG TTT TAT TTC CAA CTT TG
V_LR 3	ACG TTT CAG CTC CAG CTT GG
V_LR 4	ACC TTG GAC AGT CAG TTT GG
RS VHFor mix	*Equimolar mixture of primers RS VHF 1 and RS VHF 2*
RS V_HF 1[a]	GC**G GCC CAG CCG GCC** ATG GCC GAG GTB CAG CTB CAG CAG TC
RS V_HF 2[a]	GC**G GCC CAG CCG GCC**ATG GCC CAG GTG CAG CTG AAG SAR TC
Lin VHRev mix	*Equimolar mixture of primers Lin VHR 1, Lin VHR 2, Lin VHR 3, and Lin VHR 4*
Lin V_HR 1	GGA ACC GCC GCC ACC AGA GCC ACC ACC GCC GGA CGA GGA AAC GGT GAC CGT GGT
Lin V_HR 2	GGA ACC GCC GCC ACC AGA GCC ACC ACC GCC GGA CGA GGA GAC TGT GAG AGT GGT
Lin V_HR 3	GGA ACC GCC GCC ACC AGA GCC ACC ACC GCC GGA CGC AGA GAC AGT GAC CAG AGT
Lin V_HR 4	GGA ACC GCC GCC ACC AGA GCC ACC ACC GCC GGA CGA GGA GAC GGT GAC TGA GGT
Lin VLFor mix	*Equimolar mixture of primers Lin VLF 1 and Lin VLF 2*
Lin V_LF 1	TCT GGT GGC GGC GGT TCC GGT GGC GGT GGC GAY ATC CAG CTG ACT CAG CC
Lin V_LF 2	TCT GGT GGC GGC GGT TCC GGT GGC GGT GGC GAY ATT GTT CTC WCC CAG TC
RS VLRev mix	*Equimolar mixture of primers RS VLR 1 , RS VLR 2 , RS VLR 3, and RS VLR 4*
RS V_LR 1[b]	GAG TCA TTC T**GC GGC CGC** ACG TTT KAT TTC CAG CTT GG
RS V_LR 2[b]	GAG TCA TTC T**GC GGC CGC** ACG TTT TAT TTC CAA CTT TG
RS V_LR 3[b]	GAG TCA TTC T**GC GGC CGC**ACG TTT CAG CTC CAG CTT GG
RS V_LR 4[b]	GAG TCA TTC T**GC GGC CGC** ACC TTG GAC AGT CAG TTT GG

a Underlined bold bases indicate *Sfi*I site
b Underlined bold bases indicate *Not*I site

Reaction mix:

10× PCR Buffer:	2.0 μl
$MgCl_2$ (25 mM):	1.6 μl
dNTP (10 mM):	0.4 μl
Taq DNA polymerase (5 U/μl):	0.2 μl
Sterile nuclease-free water:	to make up the volume to 8.0 μl

4. Perform PCR amplification using a thermal cycler using the protocol given below.
   PCR amplification condition:

35 cycles	Denaturation:	95 °C/1 min.
	Primer annealing:	55 °C/1 min.
	Extension:	72 °C/2 min.
Final extension:		72 °C/10 min.

5. Resolve the PCR amplified V_H and V_L fragments on 1.5 % agarose gel (Fig. 1A). Perform gel extraction to elute the V_H and V_L fragments using a QIAquickGel Extraction Kit (*see* **Note 5**). After gel extraction, quantify the purified V_H and V_L fragments by comparing the band intensity of the purified product with that of a V_H marker (~370 bp size, 5 ng/μl concentration).

3.4 Construction of Single Chain Variable Fragment (ScFv)

Construct the ScFv gene using overlap extension PCR. Overlap extension PCR can join two or more DNA fragments into a single unit.

1. Join the V_H and V_L fragments (amplified and purified above), which have linker overhangs, via overlap extension PCR to generate ScFv fragments.
2. Use RS V_H For mix and RS V_L Rev mix for the ScFv construction (Table 1).
3. To construct a full length ScFv gene, mix equal molar ratios of V_H and V_L fragments and amplify by PCR.
4. Use hot start PCR for the construction of the ScFv. For 20 μl of reaction volume, prepare 12 μl of sample mix. Heat the sample mix at 94 °C for 5 min and then add 8 μl of reaction mix and carry out the rest of the PCR cycles as given below. The composition of the sample mix and reaction mix are given below:

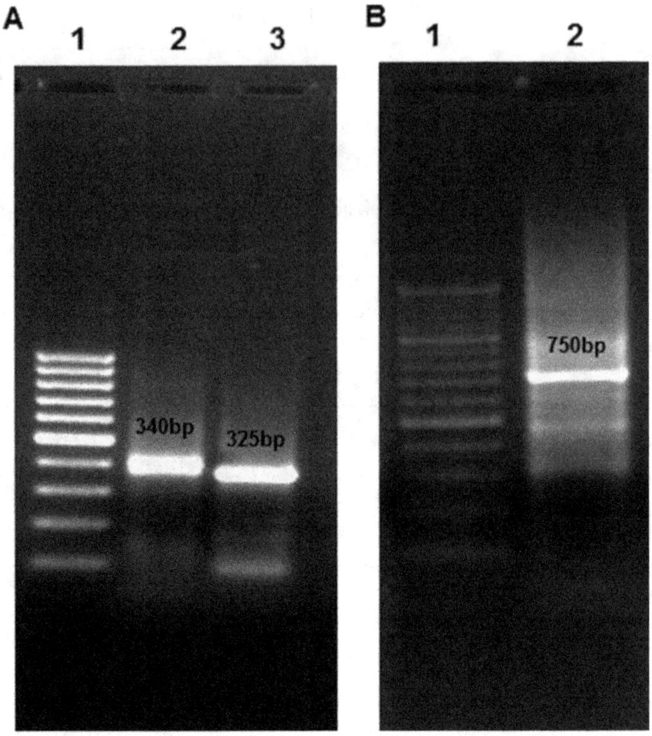

Fig. 1 Amplification of V_H, V_L, and ScFv genes. (**A**) *Lane 1*: 100 bp DNA ladder (Fermentas); *lane 2*: V_H amplicon (~340 bp), *lane 3*: V_L amplicon (~325 bp). (**B**) *Lane 1*: 100 bp DNA ladder, *lane 2*: ScFv amplicon (reproduced from Ref. [6])

Sample mix:

V_HPCR product:	~5 ng
V_LPCR product:	~5 ng
RS V_H For mix (20 pmol):	1.0 μl
RS V_L Rev mix (20 pmol):	1.0 μl
Sterile nuclease-free water:	to make up the volume to 12.0 μl

Reaction mix:

10× PCR Buffer:	2.0 μl
$MgCl_2$ (25 mM):	1.6 μl
dNTP (10 mM):	0.4 μl
Taq DNA polymerase (5 U/μl):	0.2 μl
Sterile nuclease-free water:	to make up the volume to 8.0 μl

PCR amplification condition:

35 cycles	Denaturation:	95 °C/1 min
	Primer annealing:	63 °C/1 min
	Extension:	72 °C/2 min
Final extension:		72 °C/10 min

5. Resolve the PCR amplified ScFv fragments on 1.5 % agarose gel (Fig. 1B) and elute the fragments using a QIAquickGel Extraction Kit (*see* **Note 5**).
6. Quantify the gel purified ScFv fragments on 1.5 % agarose gel by comparing the band intensity with that of a ScFv marker (~750 bp size, 5 ng/μl concentration).

3.5 Restriction Digestion of ScFv Fragment

1. The previously assembled ScFv fragment contains *Sfi*I and *Not*I restriction sites. To clone this ScFv gene into a phagemid vector (pCANTAB 5E), perform restriction digestion of the ScFv fragment as specified below.
   *Sfi*I digestion mix:

Purified ScFv DNA:	~1 μg
10× buffer M:	8.5 μl
*Sfi*I (10 U/μl):	2.0 μl (20 U)
Sterile nuclease-free water:	to make up the volume to 85.0 μl

2. Mix the components by spinning briefly and incubate at 50 °C for 4 h in a water bath. After 4 h, spin the mixture briefly and perform *Not*I digestion at 37 °C for 4 h by mixing the following *Not*I digestion mix:

3 M NaCl:	3.6 μl
10× buffer H:	1.5 μl
*Not*I (10 U/μl):	4.0 μl (40 U)
Sterile nuclease-free water:	to make up the volume to 15.0 μl

3. Deactivate the restriction enzymes by heating the mixture at 65 °C for 10 min. Resolve the digested products on agarose gel and elute the DNA fragments using a QIAquick Gel Extraction Kit (*see* **Note 5**).
4. Quantify the gel eluted products by the gel quantitation method, as described above, using a ScFv quantitation marker.

3.6 Ligation of ScFv Gene into Phagemid Vector and Transformation of Recombinant Vector into *E. coli*

Ligate the *Sfi*I and *Not*I digested ScFv fragments into pCANTAB 5E phagemid vectors for the expression of both phage antibodies and soluble ScFv antibodies. The pCANTAB 5E phagemid vector carries a 13 amino acid long peptide tag (E tag), which is expressed as a fusion protein with ScFv at its C-terminus. This helps in the detection and purification of the ScFv antibodies. Any suitable phagemids vectors can be used for this process.

1. Prepare the ligation mix, as given below, and perform ligation at 16 °C for 14–16 h in a water bath, followed by heat inactivation of the ligase by incubating the ligation mix at 70 °C for 10 min. Ligation mix

Restriction digested ScFv fragments:	150 ng
10× T4 DNA Ligation buffer:	5.0 μl
pCANTAB 5E (50 ng/μl):	5.0 μl
ATP (10 mM):	5.0 μl
T4 DNA Ligase (5 U/μl):	1.0 μl
Sterile nuclease-free water:	to make up the volume to 50.0 μl

2. Following ligation, transform the recombinant phagemids (ScFv-pCANTAB 5E vector) in electrocompetent *E. coli* TG1 cells by electroporation.
3. Thaw ten aliquots of electrocompetent TG1 cells on ice and add 5 μl of ligation mix to each tube.
4. Mix one aliquot of TG1 cells with 10 ng of covalently closed circular plasmid DNA to calculate transformation efficiency. Use one aliquot without any DNA as a control.
5. Transfer the cells and the DNA mixture to a prechilled 0.2 cm electro-cuvette and incubate on ice for 2–3 min.
6. Perform electroporation (*see* **Note 6**) by providing one pulse for each cuvette using a GenePulser programmed at 2.5 μF capacitance, 2.5 kV voltage, and 200 Ω resistance.
7. Immediately add (*see* **Note 7**) 1 ml of pre-warmed 2× YT-G broth to the cuvettes, dispense the cells into a 2 ml centrifuge tube by pipetting, and incubate for 1 h at 37 °C with shaking.
8. Sediment the transformed cells at 1000 × *g* for 5 min and resuspend the cells in 900 μl of fresh 2× YT medium. Plate 100 μl of the cell suspension on each of nine SOBAG plates and incubate overnight at 30 °C.
9. Plate 10, 50, and 100 μl of cells transformed with covalently closed circular plasmid (to calculate the transformation efficiency) and 100 μl of untransformed TG1 cells (as a negative control) on separate SOBAG plates and incubate overnight at 30 °C.

10. On the following day, flood all of the plates containing transformants from recombinant ScFv-pCANTAB 5E phagemids with 2× YT-AG medium, scrap using sterile scrapers, and pool the cells in a new 50 ml conical centrifuge tube.
11. Use some of the scraped culture mixture (~0.5 ml) for phage rescuing and store the rest of the cultures as glycerol stocks at −80 °C.

3.7 Phage Rescuing

1. Inoculate cell scraping (0.5 ml), as collected previously, in 10 ml of 2× YT-AG broth and grow at 37 °C with shaking up to OD_{600} ~ 0.5–0.6. Add helper phage M13K07 to the culture at 5:1 moi (multiplicity of infection) (Fig. 2).
2. Determine the number of PFU of M13K07 helper phage as described below:

$$\text{Helper phage PFU} = (5\times10^{8}\,\text{cells} / A_{600}\,\text{unit})\times(\text{moi of }5)\times(A_{600})\times(\text{final volume of cells}).$$

3. Incubate the above infected culture at 37 °C, first for 30 min without shaking and then for 30 min with shaking.
4. Centrifuge the culture at 1000 ×*g* for 10 min. Decant the supernatant, which has traces of medium containing glucose.
5. Resuspend the pellet in 10 ml of 2× YT-AK medium and incubate overnight at 37 °C with shaking.

3.8 Biopanning of Phage Antibody Library

Biopanning is done to selectively capture antigen binding recombinant phage particles displaying the desired antibodies. Figure 2 is a schematic representation of the steps involved in biopanning. Prior to biopanning, precipitate the recombinant phage particles (as described below) from solution with PEG to separate them from soluble antibodies. Perform biopanning on the recombinant phage particles against the desired antigen bound to a solid support. Recombinant phages that bind to the antigen will be retained, while those that fail to bind with the antigen will be removed during subsequent wash steps. When log phase *E. coli* TG1 cells are added to the panning vessel, antigen reactive phage particles will infect the bacterial cells, which are then rescued and selected again by repeated biopanning or by plating onto SOB-AG plates for the screening of individual antigen positive clones. Multiple rounds of biopanning (4–6 rounds) are required, with decreasing antigen concentration and increasing stringency and number of wash cycles at subsequent steps, as described below:

3.8.1 PEG Precipitation of Recombinant Phage Particles

1. Centrifuge the bacterial culture from phage rescuing steps (infected overnight) at 1000 ×*g* for 10 min at 4 °C.
2. Add one fifth volume of 20 % PEG/NaCl to the culture supernatant, mix thoroughly, and incubate on ice for 1 h with occasional mixing by swirling.

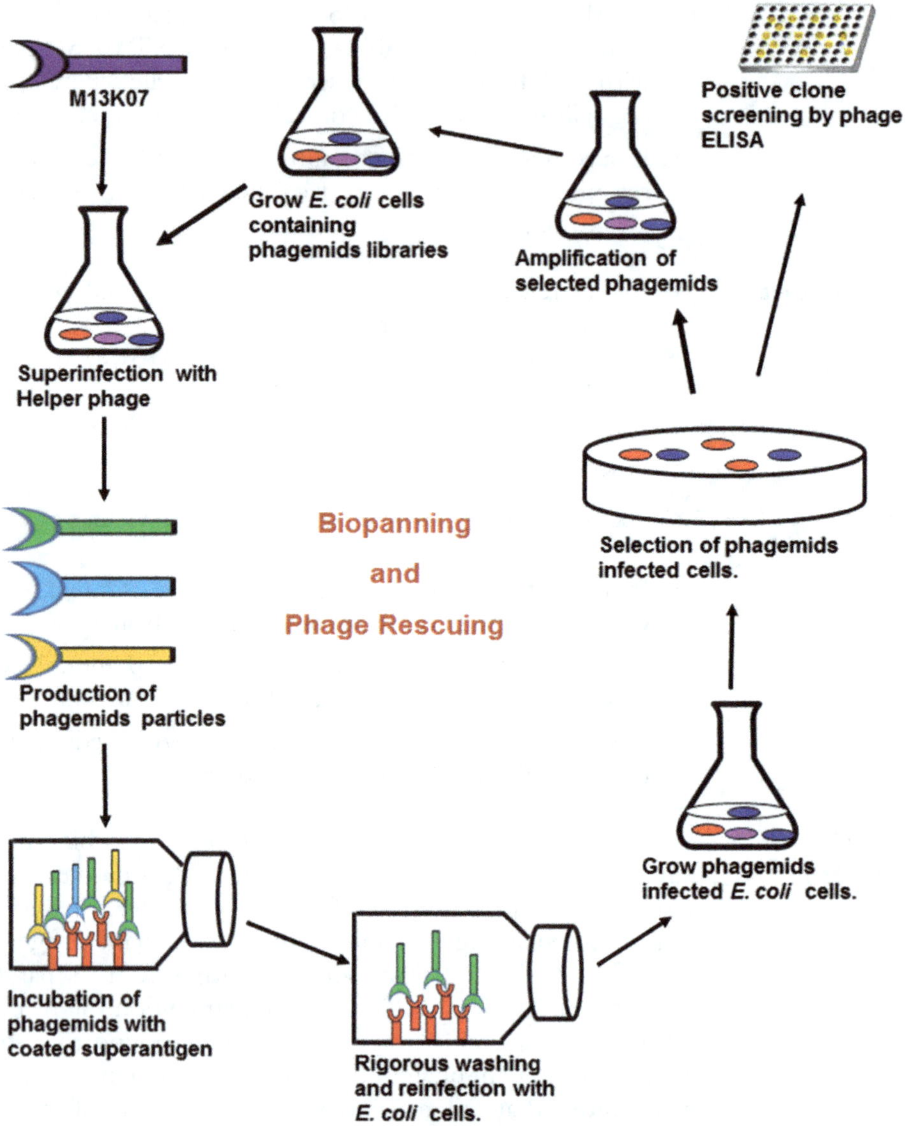

Fig. 2 Schematic representation of solid phase biopanning and phage rescuing steps involved in antibody phage display. The antibody phage library is rescued using helper phage (M13K07) and is incubated with the antigen, while the unbound phage antibodies are washed away. The bound phage particles were amplified through the infection of *E. coli* and the process is repeated to enrich the high affinity specific clones against the antigen of interest. Clones after repeated biopanning were confirmed for antibody expression by phage ELISA

3. Pellet down the recombinant phages by centrifugation at 10,000 × *g* for 20 min at 4 °C. Resuspend the phage pellet in 10 ml of 2× YT medium and store at 4 °C for immediate use.
4. Dilute the phage particles with 10 ml of 10 % MPBS and incubate at room temperature for 15 min prior to their addition to a 25 cm^2 tissue culture flask coated with the desired superantigen, as in the following panning experiment.

3.8.2 Solid Phase Panning

1. Coat a 25 cm^2 tissue culture flask with 5 ml of the desired superantigen or recombinant protein by diluting it to 10 μg/ml (*see* **Note 8**) in carbonate–bicarbonate buffer (pH 9.6) and incubate at 4 °C overnight.
2. On the following day, wash the flask three times with PBS (20 ml each), followed by blocking with 10 % MPBS for 2 h at 37 °C.
3. After blocking, first wash the flask three times with PBST and then additional three times with PBS. After this, add 20 ml of diluted phage precipitate to the flask.
4. Incubate the flask at 37 °C for 2 h and then wash five times with PBST, followed by an additional five times with PBS (*see* **Note 8**).
5. After washing, add 10 ml of log phase *E. coli* TG1 cells to the flask for reinfection and grow it for 1 h at 37 °C with shaking.
6. After incubation, remove 500 μl of culture and prepare tenfold serial dilution (1:10, 1:100, 1:1000, and 1:10,000) in 2× YT medium in different tubes. Store the remaining reinfected TG1 cultures as glycerol stocks at −80 °C.
7. Plate 100 μl of undiluted and 100 μl of each dilution of reinfected TG1 cells on SOBAG plates using a sterile glass spreader and incubate at 30 °C overnight.
8. The next day, randomly select some of the clones and patch them on fresh SOBAG plates for screening and scrape rest of the clones using 2× YT medium for the next round of phage rescuing and biopanning, as described previously.
9. After 4–6 rounds of biopanning, screen the randomly selected clones for antigen binding.

3.9 Screening of Antigen Binding Clones by Phage ELISA

Screen the antigen binding clones by phage ELISA. Produce phage particles from randomly selected clones of different biopanning steps and screen for the best antigen binders, as described below.

3.9.1 Production of Phage Particles for ELISA

1. Inoculate the randomly selected clones from the biopanning steps in 2× YT-AG medium and grow overnight at 37 °C.
2. On the following day, inoculate 100 μl of culture (grown overnight) in 5 ml of 2× YT-AG medium and grow at 37 °C with shaking up to OD_{600} ~ 0.8.
3. Add helper phage M13K07 (10^{10} PFU) to the cultures and incubate the infected cultures at 37 °C for 30 min without shaking, followed by 30 min with shaking.
4. Centrifuge the cultures at 1000 × *g* for 10 min, decant the supernatant, and remove any traces of medium.
5. Resuspend the pellet in 2× YT-AK medium and incubate overnight at 37 °C with shaking.

6. The next day, centrifuge the cultures at 1000 × *g* for 20 min at 4 °C. Collect the supernatant in a new tube, add one fifth volume of 20 % PEG/NaCl to the culture supernatant, mix well, and keep on ice for 1 h to precipitate the phage particles.
7. Centrifuge the precipitated phage particles at 10,000 × *g* for 20 min at 4 °C and resuspend the pellet in 1 ml of 2 % MPBS for use in phage ELISA.

3.9.2 Phage ELISA

1. Coat Maxisorp ELISA plates with 1 μg/well (in triplicates) of the desired superantigen or recombinant protein in bicarbonate buffer (pH 9.6) by incubating the plate overnight at 4 °C. Coat similar wells with either PBS or bicarbonate buffer for negative control.
2. Block the wells with 5 % MPBS at 37 °C for 2 h and then wash three times with PBST.
3. Add 100 μl of the precipitated phage particles from the previous section to the test antigen coated wells and negative control wells.
4. Following the addition of the phage particles, incubate the plate at 37 °C for 1 h and then wash three times with PBST. Bound phage particles will be detected by incubation with a 1:2000 dilution of anti-M13 HRP conjugated antibody for 1 h at 37 °C. After incubation, wash the plate three times with PBST and develop the color using 100 μl of TMB/H_2O_2 substrate by incubation at 37 °C for 20–25 min in the dark.
5. Stop the reaction by adding 50 μl of 2 N sulfuric acid (H_2SO_4). The plate should be read at 450/540 nm using an ELISA plate reader. Clones should be considered positive if they demonstrate at least twice the signal developed in the negative control.

3.10 Expression of Soluble Antibody Fragments

To produce ScFv antibodies in the soluble form, change the *E. coli* host strain from a suppressor strain (*E. coli* TG1) to a non-suppressor strain (*E. coli* HB2151). Reinfect *E. coli* HB2151 with the binder clones identified by phage ELISA from the previous section.

1. Rescue and precipitate the phage using PEG/NaCl from the antigen binding clones, as described in previous sections. Resuspend the precipitated phage particles in 1 ml PBS instead of 2 % MPBS and store at 4 °C for immediate use.
2. For reinfection, grow the HB2151 cells in 5 ml of 2× YT medium by incubating the culture overnight at 37 °C with shaking. The next day, inoculate 5 ml of fresh 2× YT medium with primary culture and grow the culture at 37 °C up to OD_{600} ~ 0.5 (log phase).
3. In a new tube, add 100 μl of precipitated phage tol ml of log phase HB2151 cells and incubate the cultures at 37 °C,

initially for 30 min without shaking and then for another 30 min with shaking.

4. Plate the infected HB2151 cultures on SOB-AGN plates and incubate overnight at 30 °C.
5. On the following day, pick and grid (patch) single colonies using sterile tooth picks on SOB-AGN plates.
6. Screen the clones by whole cell PCR and extract plasmids for gene sequencing to confirm the full length and correct sequence of the ScFv gene.
7. For soluble expression of the antibody fragment, inoculate 5 ml of 2× YT-AG medium with infected and uninfected HB2151 cells and grow overnight at 37 °C.
8. The next day, inoculate 2.5 ml of primary culture in 100 ml of 2× YT-AG medium and incubate at 37 °C with shaking up to OD_{600} ~ 1.0–1.2.
9. Harvest the cells by centrifugation at 1500 × *g* for 20 min and resuspend the cell pellet in 100 ml of 2× YT-A medium with 1 mM IPTG for induction.
10. After 7 h induction at 30 °C, harvest the cells by centrifugation at 1500 × *g* for 20 min at 4 °C. Resuspend the cell pellet in ice-cold 1× TES at the rate of 2 % of the initial culture volume.
11. Subsequently, add ice-cold 1/5× TES at the rate of 3 % of the initial culture volume and incubate the mixture on ice for 30 min to induce a mild osmotic shock.
12. Centrifuge the contents at 12,000 × *g* for 10 min, transfer the supernatant from the periplasm containing the soluble antibodies to a new tube, and store at −20 °C by aliquoting for further use.
13. Confirm the expression of ScFv antibody in the periplasmic extract by SDS-PAGE and Western blotting using anti-E tag antibody (Fig. 3A, B).

3.11 Antigen Binding Assay for the Recombinant ScFv

ELISA and Western blot can be used to detect the binding of recombinant antibodies to the desired superantigen.

3.11.1 ELISA

1. Coat Maxisorp ELISA plates with 250 ng per well of the desired antigen in bicarbonate buffer (pH 9.6) by incubating at 4 °C overnight.
2. The next day, wash the wells with PBST and block the wells using 300 μl of 5 % MPBS at 37 °C for 2 h.
3. After washing three times with PBST, add different dilutions of recombinant ScFv in 2 % MPBS to the antigen coated wells (100 μl/well). Periplasmic extract of untransformed HB2151

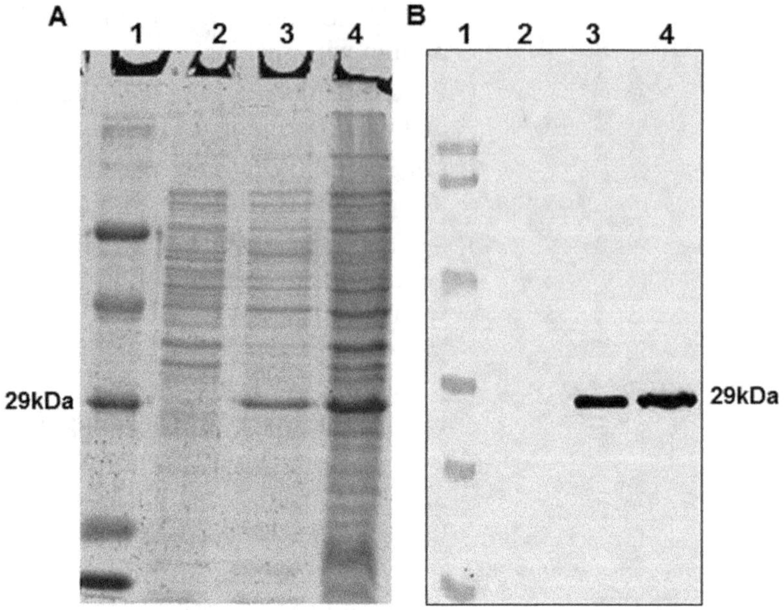

Fig. 3 Expression profile of recombinant anti-SEB ScFv. (**A**) SDS PAGE shows: *lane 1*, molecular mass markers (66, 45, 29, 20, and 14.2 kDa); lane 2, uninduced culture of 4PCL2-infected HB2151 cells; lanes 3 and 4, induced culture and periplasmic extract of 4PCL2-infected HB2151 cells, respectively, showing the anti-SEB ScFv band at ~29 kDa. (**B**) Western blot shows: *lane 1*, molecular mass markers (97.4, 66.2, 45, 31, 26.6, and 14.4 kDa); *lane 2*, uninduced culture of 4PCL2-infected HB2151 cells; *lanes 3* and *4*, induced culture and periplasmic extract of 4PCL2-infected HB2151 cells, respectively, showing the anti-SEB ScFv band at 29 kDa (reproduced from Ref. [6])

cells can be used as negative control. Incubate the plate at 37 °C for 1 h.

4. Wash the wells three times with PBST and add anti-E tag HRP conjugated antibody (1:2000) by diluting it in 2 % MPBS and incubate the plate at 37 °C for 1 h.
5. Wash the plate three times with PBST and add 100 μl of TMB/H_2O_2 substrate in each well. Incubate the plate for 20 min at 37 °C. Stop the reaction by adding 50 μl of 2 N H_2SO_4.
6. Measure the optical density at 450/540 nm using micro plate reader (Fig. 4A).

3.11.2 Western Blot

1. Resolve 1.0 μg of the desired superantigen on SDS-PAGE and transfer it to a PVDF/nitrocellulose membrane by Western blotting.
2. Block the membrane with 5 % MPBS (overnight at 4 °C or 1 h at room temperature), wash the blot three times with PBST,

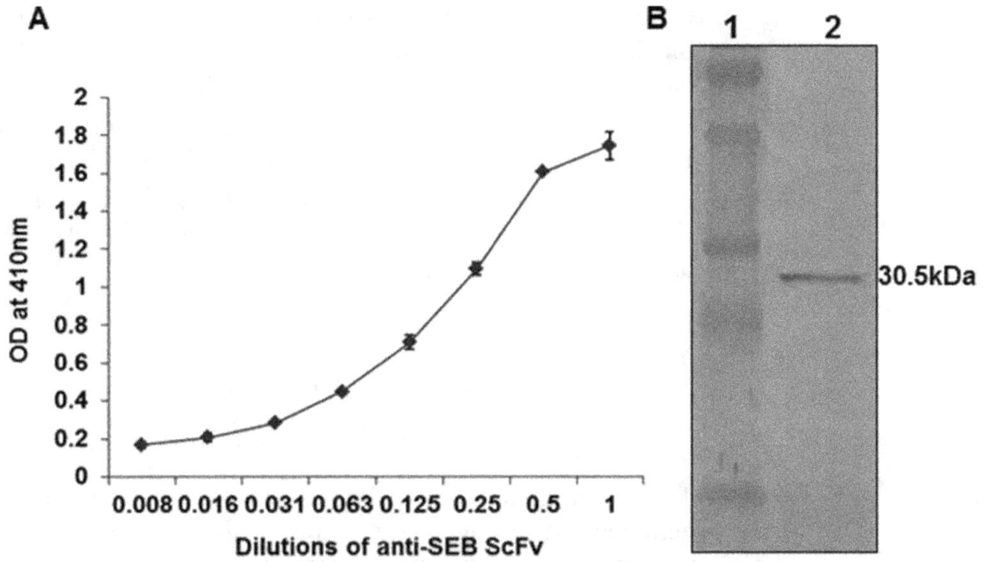

Fig. 4 Antigen binding assay of anti-SEB ScFv antibody. (**A**) ELISA was performed with different dilutions of soluble anti-SEB ScFv antibody. Periplasmic extract of untransformed HB2151 cells was used as a negative control (OD, 0.0945). (**B**) Western blot shows: *lane 1*, prestained molecular mass markers (70, 55, 35, 27, and 15 kDa); *lane 2*, r-SEB band at 30.5 kDa, showing binding with anti-SEB ScFv. The bound anti-SEB ScFv antibody was detected using anti-E-tag HRP-conjugated antibody (reproduced from Ref. [6])

and incubate it with recombinant ScFv antibody for 1 h at room temperature or overnight at 4 °C.

3. Wash the blot three times with PBST and incubate the blot with anti-E tag HRP conjugated antibody (1:2000 dilutions) for 1 h at room temperature.
4. Wash the blot three times with PBST and develop using ECL substrates by chemiluminescence or using a DAB/H_2O_2 solution (Fig. 4B).

4 Notes

1. Always check the antibody titers in the hybridoma culture supernatant and in the immunized mice serum because some hybridomas tend to lose their antibody secreting ability with time, and primed animals antibody titers have also been found to be reduced with time.
2. While pipetting upper aqueous layer always use lower volume micro tips (e.g., 200 μl). Larger volume micro tips generate unwanted force which may disturb the meniscus and mix the contaminating layers. It is always better to leave some amount of aqueous layer rather than pipetting the other contaminating layers.

3. Completely drain the ethanol from the tubes by inverting them on Kimwipes. Air-dry the tubes to evaporate the ethanol completely, any traces of ethanol will affect the solubility of RNA.
4. If not ready to process further in the following days, store RNA long term as a pellet at −80 °C in ethanol.
5. At the elution step using a QIAquickGel Extraction Kit, lukewarm nuclease-free water or EB (10 mM Tris–HCl, pH = 8.5) can be used to improve the elution efficiency.
6. Wipe out any traces of liquid on the cuvette surface to avoid sparks.
7. Minimize the time between shock and the addition of the pre-warm medium in the cuvettes. Any delay may result in cell death, leading to a lower number of clones.
8. The washing steps in the initial biopanning stages should be gentle and the antigen concentration should be high. Increase washing stringency and repetition cycles while decreasing concentration of antigens at subsequent biopanning stages.

Acknowledgement

We are grateful to the Director, DRDE, Gwalior for providing the research facilities. We thank Bruce Rottmann (Wayne State University) for critical reading and editing of this chapter.

References

1. Krakauer T (2013) Update on staphylococcal superantigen-induced signaling pathways and therapeutic interventions. Toxins 5:1629–1654
2. Miethke T, Wahl C, Heeg K, Echtenacher B, Krammer PH et al (1992) T cell-mediated lethal shock triggered in mice by the superantigen staphylococcal enterotoxin B: critical role of tumor necrosis factor. J Exp Med 175:91–98
3. Dellabona P, Peccoud J, Kappler J, Marrack P, Benoist C et al (1990) Superantigens interact with MHC class II molecules outside of the antigen groove. Cell 62:1115–1121
4. Seth A, Stern LJ, Ottenhoff TH, Engel I, Owen MJ et al (1994) Binary and ternary complexes between T-cell receptor, class II MHC and superantigen in vitro. Nature 369:324–327
5. Proft T, Fraser JD (2003) Bacterial superantigens. Clin Exp Immunol 133:299–306
6. Singh PK, Agrawal R, Kamboj DV, Gupta G, Boopathi M et al (2010) Construction of a single-chain variable-fragment antibody against the superantigen Staphylococcal enterotoxin B. Appl Environ Microbiol 76:8184–8191
7. Fast DJ, Schlievert PM, Nelson RD (1989) Toxic shock syndrome-associated staphylococcal and streptococcal pyrogenic toxins are potent inducers of tumor necrosis factor production. Infect Immun 57:291–294
8. Neumann B, Engelhardt B, Wagner H, Holzmann B (1997) Induction of acute inflammatory lung injury by staphylococcal enterotoxin B. J Immunol 158:1862–1871
9. Vial T, Descotes J (1995) Immune-mediated side-effects of cytokines in humans. Toxicology 105:31–57
10. Frame KK, Hu WS (1990) The loss of antibody productivity in continuous culture of hybridoma cells. Biotechnol Bioeng 35:469–476
11. Kessler N, Bertrand S, Aymard M (1993) Stability of a murine hybridoma is dependent on the clonal line and culture media. In Vitro Cell Dev Biol 29A:203–207
12. Ozturk SS, Palsson BO (1990) Loss of antibody productivity during long-term cultivation

of a hybridoma cell line in low serum and serum-free media. Hybridoma 9:167–175

13. Skerra A, Pluckthun A (1988) Assembly of a functional immunoglobulin Fv fragment in *Escherichia coli*. Science 240:1038–1041
14. Ho M, Nagata S, Pastan I (2006) Isolation of anti-CD22 Fv with high affinity by Fv display on human cells. Proc Natl Acad Sci U S A 103:9637–9642
15. Galeffi P, Lombardi A, Pietraforte I, Novelli F, Di Donato M et al (2006) Functional expression of a single-chain antibody to ErbB-2 in plants and cell-free systems. J Transl Med 4:39
16. Choo AB, Dunn RD, Broady KW, Raison RL (2002) Soluble expression of a functional recombinant cytolytic immunotoxin in insect cells. Protein Expr Purif 24:338–347
17. Coia G, Hudson PJ, Irving RA (2001) Protein affinity maturation in vivo using *E. coli* mutator cells. J Immunol Methods 251:187–193
18. Irving RA, Kortt AA, Hudson PJ (1996) Affinity maturation of recombinant antibodies using *E. coli* mutator cells. Immunotechnology 2:127–143
19. Wang SH, Zhang JB, Zhang ZP, Zhou YF, Yang RF et al (2006) Construction of single chain variable fragment (ScFv) and BiscFv-alkaline phosphatase fusion protein for detection of Bacillus anthracis. Anal Chem 78:997–1004
20. Sato M, Kojima K, Sakuma C, Murakami M, Tamada Y et al (2014) Production of scFv-conjugated affinity silk film and its application to a novel enzyme-linked immunosorbent assay. Sci Rep 4:4080
21. Kim H, Park S, Lee HK, Chung J (2013) Application of bispecific antibody against antigen and hapten for immunodetection and immunopurification. Exp Mol Med 45:e43

INDEX

Amanda J. Brosnahan (ed.), *Superantigens: Methods and Protocols*, Methods in Molecular Biology, vol. 1396, DOI 10.1007/978-1-4939-3344-0, © Springer Science+Business Media New York 2016

www.ingramcontent.com/pod-product-compliance
Lightning Source LLC
LaVergne TN
LVHW081000280926
849069LV00013B/554

* 9 7 8 1 4 9 3 9 8 0 2 6 0 *